Commission of the European Communities

Information Symposium

COKE OVEN TECHNIQUES

Luxembourg, 23 and 24 September 1981

The three language versions of these proceedings are being published in partnership by:
— Verlag Glückauf GmbH, Essen (Federal Republic of Germany), German language edition
— Graham and Trotman Limited, London (United Kingdom), English language edition
— Société de l'industrie minérale, Saint-Etienne (France), French language edition

GRAHAM & TROTMAN LIMITED

EUR 7808 d, e, f,

Organisation of the Information Symposium
by
The Commission of the European Communities
Directorate-General Energy
and
Directorate-General for Information Market and Innovation

Publication arrangements:
P.P. ROTONDO'

LEGAL NOTICE

Published by:

Graham & Trotman Limited, Bond Street House
14 Clifford Street London, W1X 1RD.

ISBN-13: 978-0-86010-366-0 e-ISBN-13: 978-94-009-7367-1
DOI: 10.1007/978-94-009-7367-1

CONTENTS

OPENING SESSION

- Opening address

  C. POLFER, Mayor of the city of Luxembourg

- Introductory address

  Coke supply and demand in the European Community

  K. REICHERT, Director for coal in the Directorate General for Energy, Commission of the European Communities

- General introduction

  Coke and coal in the changing technology of the blast furnace

  G. SIRONI, Director, Centro Sperimentale Metallurgico

OPENING ADDRESS BY
Mr. C. POLFER
Mayor of Luxembourg

Ladies and gentleman,

When I received your kind invitation to give an opening address at your symposium on coke oven techniques, I could not help thinking, at a time when my country is in a difficult situation in the European Community, of the beginnings of this long and apparently unending process of European unification. It was here, in our little capital city, that this process began in 1952 with what at that time was called the ECSC, now the Commission of the European Communities.

That genius Jean Monnet wrote in his memoirs, "On the morning of 10 August 1952 all Europe was in Luxembourg. The charming capital of the Grand Duchy was not prepared for this influx of people of different origins and with different attitudes who without wasting a second settled down to working at speed amid a more stolid people. Within a few days venerable government departments were transferred to make room for our new institutions, the first units of which were arriving piecemeal in waves from Bonn, Paris, Rome, Brussels and the Hague, like invaders from another world".

Since that time these "invaders from another world" and the "stolid" people of Luxembourg have travelled a fair distance along the road together, with both sides learning to recognize and appreciate the other's qualities. Today Luxembourg is no longer the rather sleepy little old town which is was just after the war and which hastily moved out its own government departments to make room for the exponents of a concept of Europe which it had supported fervently from the outset.

Luxembourg has always been, geographically, one of the crossroads of Europe, and it has become one of the crossroads of the European Community as well. It has expended all its energies and resources in the service of the European cause, and over the years has managed to make the Kirchberg a real European centre. And now, at the end of this road, along which there have been so many sacrifices, we find that some people want to wipe out our capital city without more ado, like some wicked fairy waving a magic wand.

I realize that I am speaking to an assembly of eminent experts who are about to spend 2 days discussing the technical problems of coking plants and the coal industry in the Community. While I am not an expert on coke-making I am, on the other hand, a convinced European, and that is why, as mayor of this city, I was anxious to stress a matter which is particularly close to my heart.

I fully realize the importance of your work, particularly as it concerns Europe's crisis-ridden steel industry, and you will not be unaware that Luxembourg has its own problems in this area as well. The field you will be concerned with here is of crucial importance in the utilization of coal; coke is a valuable outlet for the Community's coalmining industry and is fundamental to the manufacture of steel products.

The steel industry is still a prime consumer of Community coal, and will remain so for the next 20 years. The forecasts are that steel production will rise to 160 - 170 million tonnes in 1990 and to 180 - 190 million tonnes in the year 2000. At present more than 40% of Community coal production is turned into coke. Productivity at the blast furnace is a function of coke quality, and consequently of coke-making techniques.

In order to provide a regular review of the state of this technology the European Community organizes symposia from time to time to pass on information. This symposium opening today is the third, following those held in 1970 and 1975.

Ladies and gentlemen, I hope that these two days spent in our capital will bring you a wealth of information and results which cannot fail to be beneficial for the future of the European steel industry. I also hope that your stay with us, which is all too brief, will be as enjoyable as possible and that you will retain an excellent impression of your visit here. Lastly, I hope that I shall have the pleasure of welcoming you here again in the not too distant future, when perhaps the clouds now darkening the European sky will have cleared away and the crisis which is troubling us now will be no more than an unhappy memory.

Coke supply and demand in the European Community

by Dr. Karlheinz Reichert, Director for Coal,
Commission of the European Communities

Ladies and Gentlemen,

On behalf of the Commission of the European Communities I have great pleasure in welcoming you to Luxembourg and to our Information Symposium on Coke Oven Techniques. I have been asked particularly to pass on greetings from Mr. Davignon, Vice-President of the Commission with special responsibility for coal and steel. We are delighted that there has been such a big response to the Symposium and we especially welcome delegates from non-Community countries who have not been deterred by the long distances they have had to travel. I wish you all a pleasant and interesting stay.

We have tried to arrange the programme for these two days in such a way as to ensure that all of you, whether you are researchers, scientists or on the operational side of the industry, return from Luxembourg with an equally instructive insight into what progress has been made in coking technology and related fields. This is the third conference of this kind that we have held, the previous ones having been in 1970 and 1975.

Coking is still one of the major themes of research in projects promoted by the Commission through the ECSC budget because of its importance to steel-making and because, since the enormous increase in the cost of all forms of energy, coke has become the single most expensive item in pig-iron production. This is more than enough to explain why, in view of the efforts to put the Community steel industry back on a competitive footing, we must continue to improve the technical performance and the economics of coking plants.

But, before the technical papers, let us take the opportunity to review the market for coke or, to be more precise, the Community market for metallurgical coke, as it is some time since this aspect was covered at one of these conferences. It should be clearly understood that what follows is strictly limited to the Community and the coke sector. Please do not look for any forecasts - whether for steel or for coking coal. The rather more detailed discussion of certain technical aspects is intended to compensate you for our refraining from crystal-ball gazing.

1.

Experts are apparently agreed that smelting in a blast furnace is still the best process for producing iron as a preliminary to making steel in a converter. Advances since the war - larger output, improved quality, greater uniformity, more efficient use of energy - have helped the blast furnace to maintain its traditional position; in other words: compared with other processes, and particularly in these days of the mass production of steel, smelting iron ore in a blast furnace has remained competitive. More than half the steel produced worldwide is made from blast-furnace iron in converters. In the Community and the big steel producing countries over 70% is produced by this method and the proportion is likely to grow as the open-hearth method fades away.

Direct reduction of iron ore, a process whose potential has been discussed in the past, is unlikely to enjoy a wide application as there is no source of cheap gas in the Community.

It is impossible to run a blast furnace without coke. Coke is a reducing agent, energy source, means of support and carburizing agent all in one. This brings me to my subject - namely, coke.

Fig. 1 At the moment hard-coal coke consumption in the Community is just over 60 million tonnes. This is the "market framework" within which we have to work and consists, roughly, of five separate or sub-markets:

1. the steel industry, which uses mainly blast-furnace coke but also some other types, e.g. for sintering iron ore;

2. the rest of industry, such as chemicals, electrometallurgy, iron and non-ferrous foundries, etc.;

3. households, which use coke for heating;

4. all other consumers; this covers a very wide range;

5. exports to non-Community countries.

Over the years all these markets except the export market have declined in relation to the steel industry. In 1960 the steel industry took some 63% of coke production; in 1980 it was 84%. This has meant that the utilization of coking capacity has become more dependent on the prosperity of the steel industry. Because of the significant decrease in the size of the other markets, relatively higher prices obtainable there can no longer compensate sufficiently when sales to the steel industry decline during a recession.

Fig. 2 Unfortunately, after decades of unparalleled prosperity in pig iron and crude steel up to 1974, it is not at the moment a question of recession but of a steel crisis. The first oil-price crisis was on us in 1974, of course. Because the Community's energy supply problems are known to us all I do not need to explain in detail why the energy-intensive steel industry has run into difficulties worldwide but most particularly in Europe.

Figure 2 shows how the coke rate (per tonne of pig iron) dropped as a result of greater use of high-grade ores, sinter and pellets, larger blast furnaces and oil injection. For years it was thought that the lower limit was still a long way off. But since the second oil-price crisis in 1979 we have noticed that the coke rate is beginning to rise again owing to a further reduction in the rate of oil injection. High-performance blast furnaces are again being operated without oil.

In other words, the entire pattern of energy use in the coke-based process route is moving towards the exclusive use of solids: coal, coke and coke breeze[1].

Fig. 3 The next Figure illustrates the long-standing system of supplies to the Community's coking plants. We can distinguish three main sources: domestic coal (i.e. the coking coal produced in the country in question, be it Belgium, Germany, France or the United Kingdom); coal from other Community countries; and coal imported from non-Community countries, mainly the United States, Australia and Poland. We note first that coal input has dropped from 120 to 85 million tonnes. In addition the proportion of Community coal has decreased from 88 to 72%. As the governments of coal-producing Member States have to provide heavy subsidies for the production and marketing of their coal, intra-Community trade in coking coal has decreased slightly, to the benefit of imported coal.

Figure 3 illustrates the input side, in other words it deals with coal, not coke. Here we must mention a fact to which we will be returning. There is no established world coke market - at least not for blast-furnace coke. But since the Community - unlike the rest of the world - has many coking plants run by mines and a number of independent coking plants, we in Europe are an exception in that we have a market in blast-furnace coke. I do not intend to go into the reasons here. However, the facts are that, besides the intra-Community flows of coking coal (Figure 3) there are, by tradition, further flows of metallurgical coke currently amounting to some 6 to 8 million tonnes, most of which is supplied by Germany to her neighbours, Belgium, France, Luxembourg and the Netherlands, but some of which is traded between the latter countries. In this sense we may talk of a market that developed even before the ECSC was established. The Community's blast furnaces, particularly those on the Continent, to a far greater extent than is possible elsewhere in the world, can rely for supplies on coking plants other than their own in-plant ones,

---

[1] "Die Weltenergiesituation und ihre Auswirkung auf die Eisen- und Stahlindustrie" (The world energy situation and its effects on the iron and steel industry) by Aichinger, Hoffmann, and Pittel, published in Stahl und Eisen (101), 1 July 1981, pp. 181/193

and do so – some of them extensively. A special example is the Luxembourg steel industry which gets all its coke from the three pit-head coking plants it runs at mines situated in Germany.

2.

Nowadays when the subject of steel policy comes up, discussion always centres on the reduction of capacity, on modernization and on rationalization. When it is the turn of coal policy the theme is investment and, above all, conversion of the electricity-supply industry, and industry in general, from oil to coal. The question of subsidies is discussed in connection with both coal and steel, but coke, which has to do with both coal and steel, is not much discussed. Yet this industry is a link between coal and steel and so of major importance for the future of both.

In the last 12 months the Commission's Coal Directorate has therefore conducted an opinion survey among coking-plant and blast-furnace operators in an effort to determine existing capacity, the age of coke ovens and what plans the operators have for refurbishing existing units or building new ones. With the exception of a few small independents, all Community coking plants were included in the survey; the great majority of those surveyed tried to answer all the questions asked. Unfortunately, in a few important instances the answers received were incomplete. Generally speaking questions relating to the more distant future (beyond 1990) were, understandably, answered with extreme reticence. In what follows we will give you a summary of the results. All figures are for 1980.

For the sake of completeness and clarity it should be noted that, although Greece has been a member of the European Community since the beginning of this year, it is not included in this survey. From the point of view of understanding the problem as a whole this is not serious, as Greece imports less than a million tonnes of coke a year and produces less than a million tonnes of pig iron. For our purposes these quantities are marginal.

Fig. 4 At present there are 349 coking batteries in the Community. They are concentrated in various regions as shown, somewhat diagramatically, in Figure 4; they break down further into pit-head coking plants, steelworks coking plants and independent coking plants. Their capacity is 76.5 million tonnes and the utilization rate was 83% in 1980.

Taking the situation in the world as a whole, apart from the town-gas coking plants (all of which have now been closed down) and a number of independent plants, all coking installations are run by the steel industry to feed its blast furnaces. For special, historical reasons which cannot be discussed here in detail there are many pit-head coking plants in the Community, particularly in Germany, France and the United Kingdom. These pit-head plants have a current capacity of 32.5 million tonnes: no less than 42% of total coking capacity.

Fig. 5 The age distribution within the various categories is as follows:

Of the coking plants at coastal steelworks 90% are less than 20 years old and 60% are less than 10 years old. By contrast, almost two-thirds of the coking plants at inland steelworks are over 20 years old and only 28% are less than 10 years old.

The situation with pit-head and the relatively few independently run plants is similar, 78% being over 10 years old and 61 % over 20 years old.

Later, when we discuss investment intentions, we shall be returning to the age structure.

3.

Of course, the future is always more interesting than the past or present. How will the demand for coke develop and will there be enough coking capacity? The first is a question about the medium- to long-term future of the Community's steel industry, whilst the second seeks to discover whether we can count on a quantitative balance between supply and demand.

I said at the outset that you must not expect any forecasts. Who at present would dare to forecast the steel industry's future convincingly and with reliable figures? What follows uses figures in a tentative effort to answer these questions, but is a mere substitute for a proper forecast.

In the last ten years there has been no dearth of forecasts of the economic future of the Community, its individual industries and the energy sector. Time has passed and for obvious reasons all the forecasts have been revised downwards - several times in some cases. There is at present a distinct world-wide lack of useful forward estimates. What we need is some idea of future levels of pig-iron production so that we can calculate the future demand for coke.

In the past a figure of 945 million ingot tonnes by 1980, the estimate put forward by the UN Economic Commission for Europe at Rome in 1973, was frequently quoted for the world steel industry. We know that this was too high by one-quarter, the actual figure being 719 million tonnes; similarly with forecasts of demand for coal and coke.

The Commission of the European Communities did not do much better in its "General Objectives: Steel". As compared with a forecast of 134 to 140 million ingot tonnes in 1980, the Community actually produced only 128 million ingot tonnes.

Neutral observers[1] believe that the growth trend in demand for steel will continue and say that saturation point has not yet been reached. According to them, growth has been interrupted temporarily by the rise in energy prices and attendant phenomena, with the growth curve shifted to an indeterminate point in the future at which it will resume. But even these experts cannot pinpoint when that time will be.

---

[1] "Die Grenzen des Verbrauchs und der Herstellung von Stahl in der Welt" (Limits of world steel consumption and manufacture) by Schenck, published in Stahl und Eisen (101), 1 July 1981, pp. 231/238

With things as they are one is grateful for the note of optimism. However, the steel industry has only just begun to restructure and does not yet share the optimism. In our survey we asked how each company expected its pig-iron capacity to develop and received either no answer at all or only partial answers. At any rate, the answers provided no conclusive material for the period after 1985. We shall therefore make

Fig. 6 the simplified assumption that overall capacity will remain unchanged after 1985 (135 million tonnes of pig iron). There are good reasons to dispute whether this is realistic, since it is precisely on the blast-furnace front that there will be major changes in the next few years, be they in the form of positive rationalization (modernization; concentrating production in very large blast furnaces) or negative rationalization (reducing capacity). A simplified assumption of this kind still leaves out certain factors: for example, in a steel slump, wider use is made of scrap-based steel processes and less use is made of pig-iron-based processes.

To obtain figures for future coke demand we have to make assumptions about the coke rate per tonne of pig iron produced (514 kg in 1980) and about other levels of coke consumption in the steel industry, and combine these figures with various levels of pig-iron production. Figure 6 represents these assumptions as the rate of utilization of blast-furnace capacity. Therefore in Figure 6 the actual 1980 production of 89.5 million tonnes of pig iron is translated into a rate of utilization of 65% of last year's actual capacity of 137.2 million tonnes.

Where the coke rate is concerned, it is reckoned that, following a period when a drop in consumption turned into a rise as coke replaced oil, there will be a further drop because of blast furnace modernization, closures of old units and the building of modern very large blast furnaces. These, too, are only simplified assumptions.

Having dealt with consumption, let us now look at future developments in coke supplies and coking capacity. In this area the figures can be regarded as much more reliable although we had to make some assumptions in order to obtain a full picture. The assumptions took into account the state of the coking plant in question and various other factors. Figures for new plant and rebuilds were given equal weighting in the calculations to determine future developments in capacity.

Fig. 7 Figure 7 is similar to Figure 4 but with the information presented by country (Benelux being treated as one country) in order not to reveal companies' investment plans. The following observations, based on our knowledge of individual replies, can be made in connection with this Figure:

1. Total coking capacity will decrease from 76.5 to 66.4 million tonnes per year in the next 20 years.

2. Depending on the age structure of existing coking plants, some regions and countries will be investing more and others less. In France, for instance, there are only limited plans for the more distant future. Neither does the Italian steel industry plan any major investment. About half the existing coking capacity in the Benelux countries will be replaced. In the United Kingdom it seems that only some of the coking plant reaching the end of its useful life is to be replaced or modernized. Germany is the only country in which, according to the firms' answers, there will be a slight increase in capacity, but only if the mining industry actually carries out its building programme for the next 20 years.

3. The present age structure of our coking plants, with 45% of capacity of 20 years and over, shows that new building and modernization of existing units have been neglected. One reason might be the relatively high capital costs.

4. From the information provided it appears that a total of 43 million tonnes of new capacity is to be built, 12 million by 1985, a further 18 million by 1990 and another 13 million by 2000. Coastal steelworks plan to build 5 million tonnes of new capacity whilst inland steelworks plan 10.5 million and the mining industry and the independents 27.7 million tonnes of new capacity - 23.9 million of the latter being accounted for by the German coal mining industry alone. This point merits particular attention. If for any reason the mines are unable to complete their renewal programme, the Community's total coking capacity will drop much further and faster than indicated and this could mean serious coke supply difficulties.

5. We received 16 replies, with a wide range of figures, on the capital cost of refurbishing existing coke ovens. The bracket spread is US $ 34 to 122 per tonne/year of capacity with the largest number of replies lying between US $ 50 and 85, and the second largest at US $ 115.

Nine undertakings gave information on the specific cost of new plant; their investment programmes range from 500,000 to 2.5 million tonnes annual capacity. Here the figures fluctuate between $ 80 and 250 with some bunching in the range of $ 120 to 190 per tonne of coke per year. This makes building from scratch twice as expensive as a rebuild, but we also have figures showing that to rebuild is more expensive than to build a new plant. This indicates the use of different costing methods, which explains the wide scatter.

On the other hand, one could infer that in some instances it would be more advantageous to keep on coking at an old site and forgo the optimum conditions attainable with an entirely new project because anti-pollution facilities, for example, would excessively increase the capital cost.

6. High capital costs like these are necessarily reflected in higher operating costs. Will the coking-plant revenues rise accordingly?

That depends on how the market develops, a question no one can answer at present. However, the reversal of the price relativity of blast-furnace coke and heavy fuel oil, based on calorific value, is worth mentioning. Around the end of 1973/beginning of 1974, coke was twice as expensive as fuel oil. Five years later they were equally expensive and by the summer of this year oil was 55% more expensive. And, of course, the revenue situation of the liquid and gaseous by-products of coal has changed. All this affects the costing of investment in coking plants.

5.

What may we now conclude from this juxtaposition of the assumed trend on the demand side and the possible trend of capacity?

Fig. 8 The figures in Figure 8 are based partly on Figure 6 and give simplified assumptions with regard to pig-iron capacity, the various levels of coke consumption in the steel industry - depending on blast-furnace capacity utilization - and the possible coking capacity.

Supplies to non-steel consumers must be subtracted from this maximum supply of coke. The demand of the other sectors is relatively price-inelastic; it is therefore likely that even with a market under stress this demand could be satisfied. Only the remainder is available to the steel industry. According to the information provided by coking plant operators, coking capacity is to decrease so that, presumably, less coke will be available even if coking plants are used to capacity. Only the future will tell whether this will bring supply bottlenecks or whether the margin is adequate. From the figures we have put together here we can calculate only that the available coke would permit the utilization of 77% of a pig-iron capacity of 135 million tonnes; in other words some 105 million tonnes of pig iron could be produced. Nowadays this seems like a dream come true and yet a total capacity of 66 million tonnes of coke towards the end of the century would seem to be on the small side rather than excessive.

In this connection I must repeat my earlier caveat: we did not receive any useful data from a number of key steel undertakings in reply to our questions on their likely pig-iron capacity after 1985. In Figure 8 we are juggling with assumed figures. But consider for a moment what fundamental changes there would be if investment in coking plant were not carried out on the planned scale. First, total demand and total supply would probably soon be out of phase and the estimated 77% utilization of pig-iron capacity would be a pipe-dream. The situation would be particularly precarious if mine-operated coking capacity was not renewed at all or if not enough of it was renewed. This would cause perceptible imbalance in certain regions and two alternative conclusions can be drawn:

Reasonably adequate utilization of blast-furnace capacity in the Community requires the traditional flows of coking coal to continue. In other words: the common coke market must remain intact.

Or else: The steel industry in some Community regions must abandon the historic pattern of coke supply and build its own coking capacity. Given the steel industry's present difficulties, it could not easily cope with a new departure of this kind.

6.

Ladies and Gentlemen,

I conclude with some remarks on the market for coking coal. It seems that there is no reason to expect surprises either on the demand or on the supply side. Enough coking coal for the steel industry is available. Of course there will always be a demand for prime coking coal and supplies of this commodity are to some extent limited.

The signs of change are to be found elsewhere - in the related markets for coal for power stations and for industry. Here the supply and demand situation has recently caused prices to rise sharply and has led to a distinct narrowing of the price gap between coking coal and steam coal. This can be blamed on the second round of oil price increases on the world market in 1979/80. In the wake of these increases the Community's

Council of Ministers laid down new energy policy goals for 1990. They state inter alia that consumer prices for energy must reflect representative world market conditions and the cost of developing new energy sources. The rise in the US dollar is a further important factor.

In this context a condition the Commission laid down in 1970 when it completely revised the still existant coking coal scheme gains new significance. The condition is that the scheme should make it possible for the steel industry to bear the full cost of its coke supplies. The present state of the world coal market means that the Community's coal policy - which also covers the coking-coal sector - can and must be reviewed in the context of Community energy policy and, if necessary, be given a fresh slant in various respects. This raises again the old familiar question of the relationship between price and security of supply, which is given special overtones by general developments in the energy sector. Hence the Commission does not wish to renew the coking coal scheme, which is due to expire at the end of 1981, for more than two years. An overall view of the situation is now needed; it will have to combine highly diverse elements: the restructuring of the European steel industry, which is just beginning; the security of energy supplies; the long-term objectives for the coal sector; how much of a subsidy burden national budgets can bear, which is again becoming highly topical; etc.

One cannot imagine considering this situation and yet leaving out the coke industry. The aim of this introduction was to give you some idea of the industry's present shape and what the future holds for it. From now on we shall be concerned with coking technology. And so I hereby declare this Symposium on Coke Oven Techniques open.

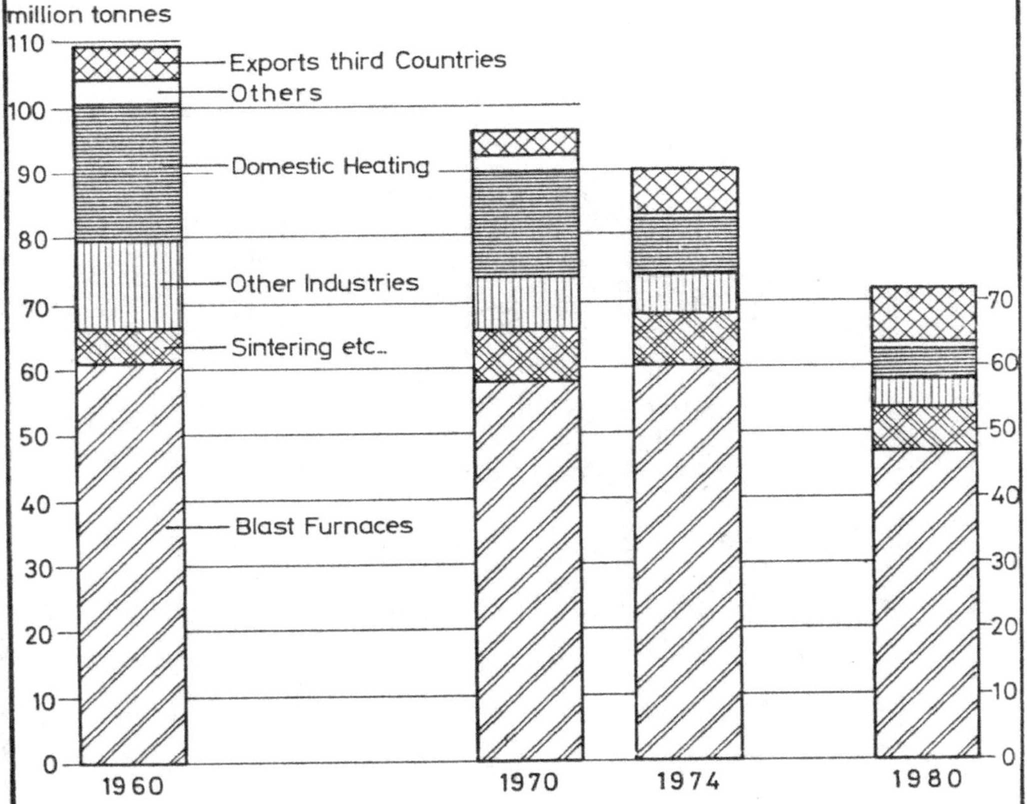

Graph.1 Coke Consumption by Sector (Eur-9)

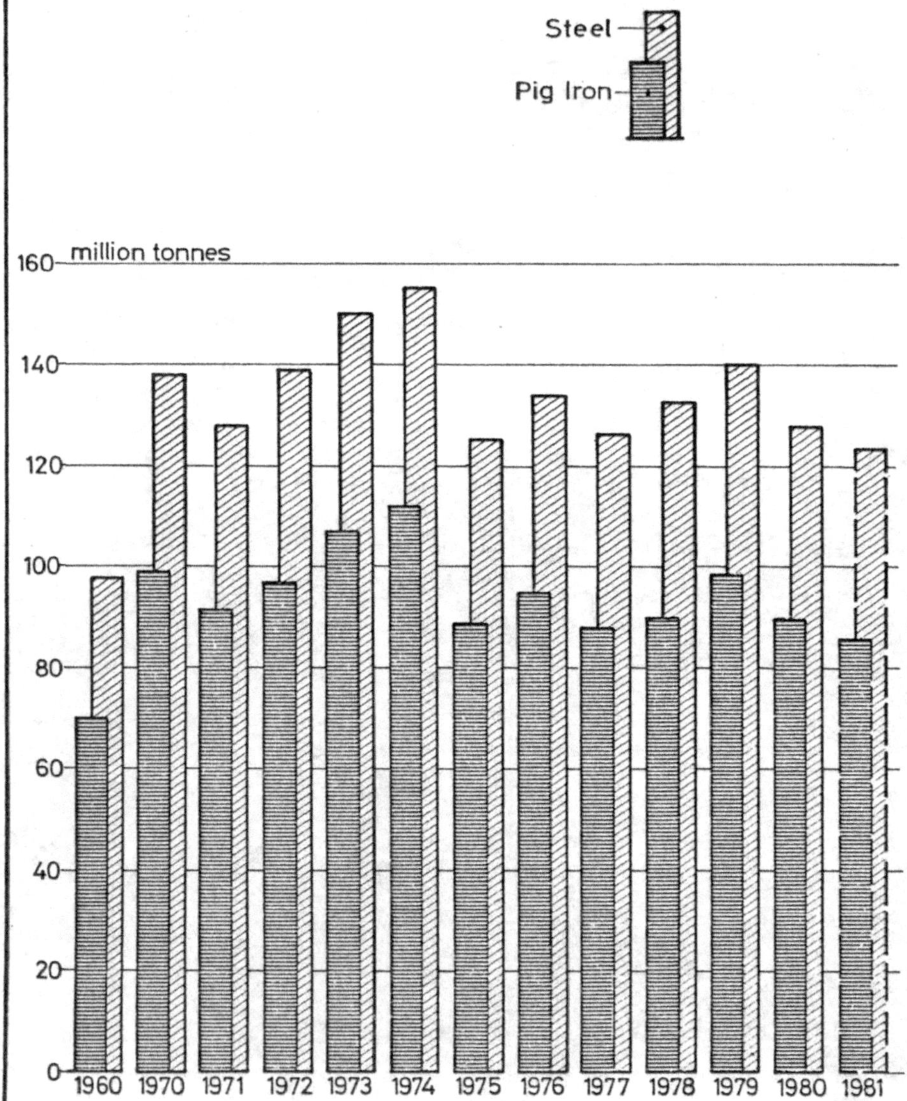

Graph.2 Pig Iron and crude Steel Production (Eur-9)

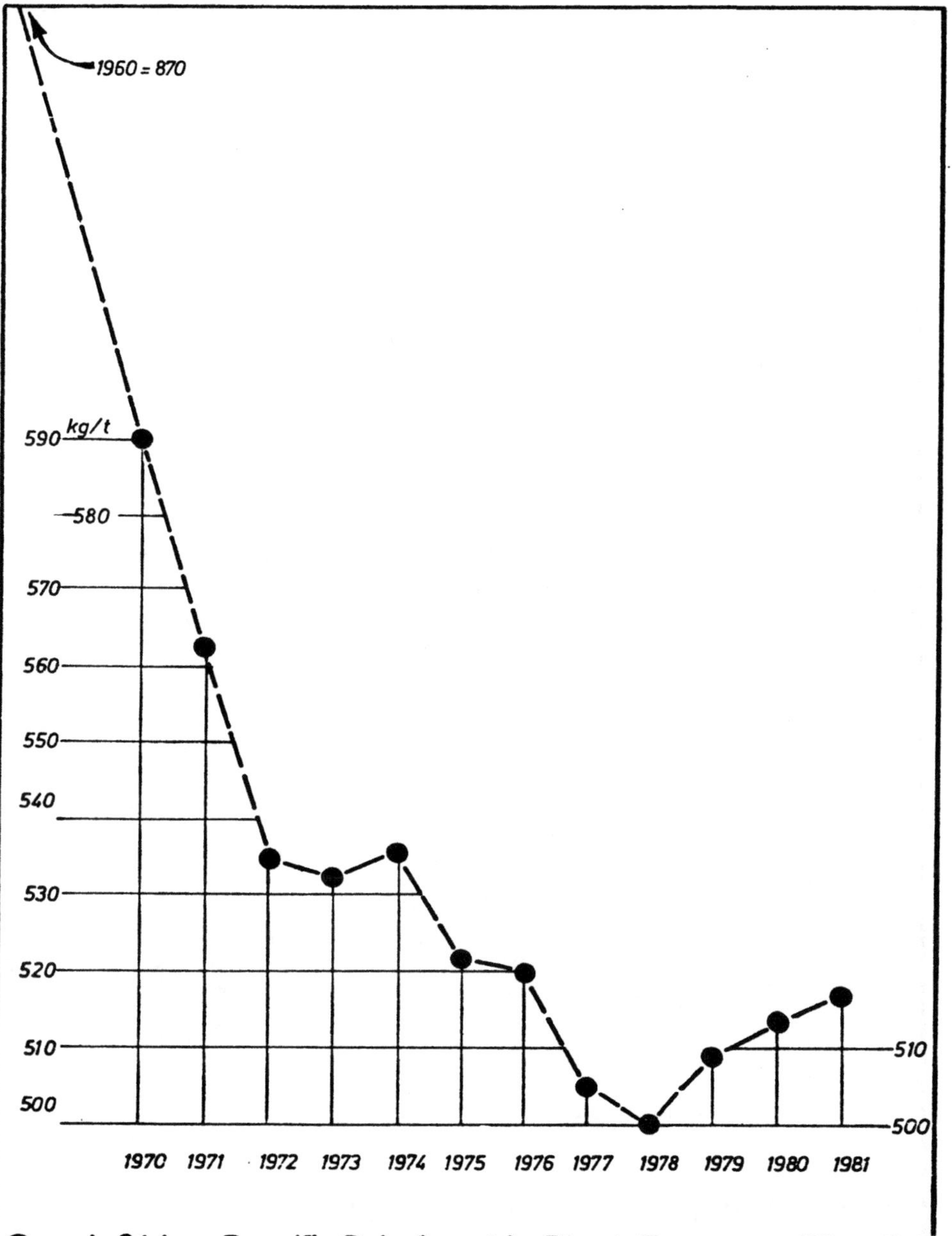

Graph. 2 bis - Specific Coke Input in Blast Furnaces (Eur- 9)

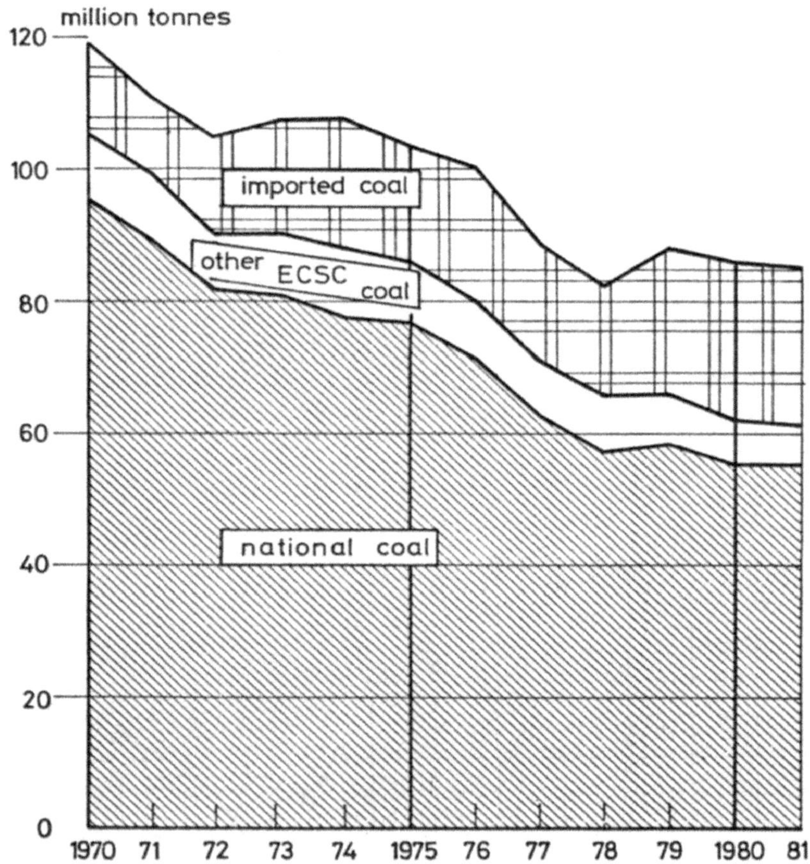

Graph. 3 Hard Coal Supply to Coke Ovens (Eur-9)

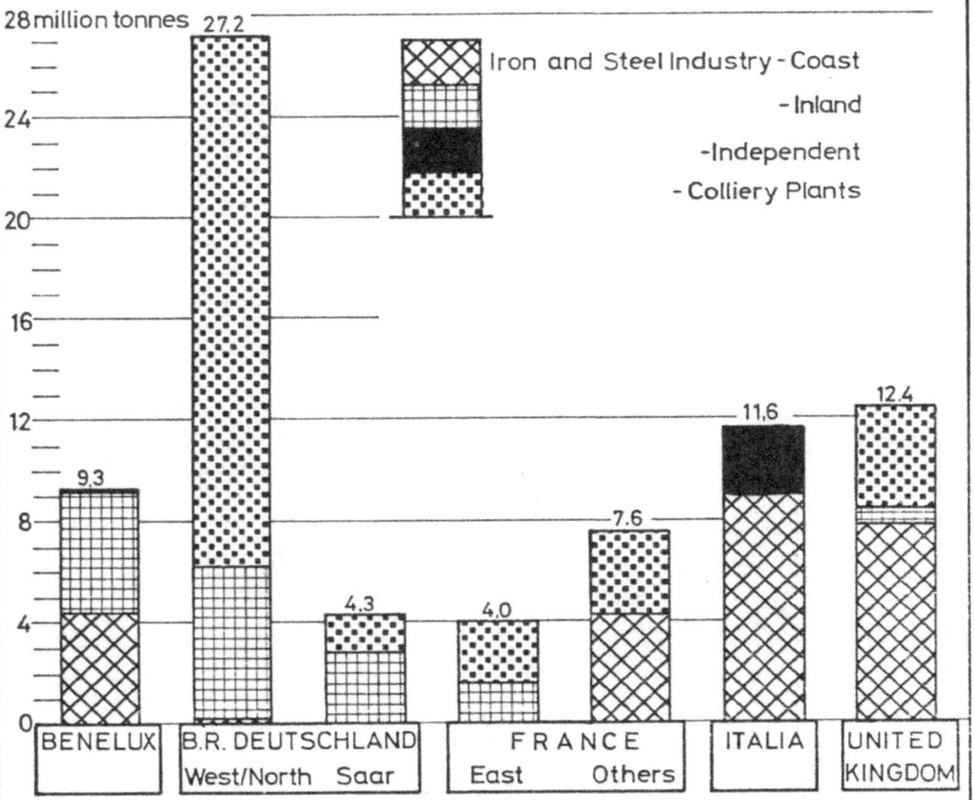

Graph. 4 Coke-Oven Capacity by Regions and Categories (Eur-9)

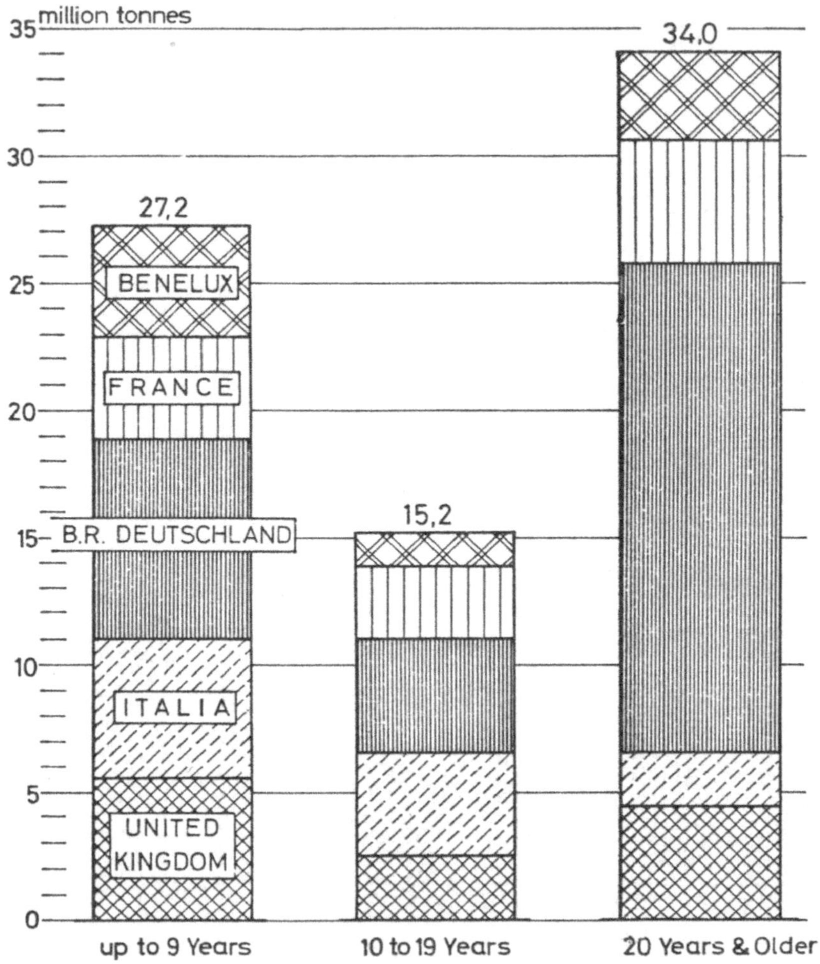

Graph.5 Age Structure of Coke-Oven Batteries in 1980 (Eur-9)

	1980 actual	1985	1990	2000
Pig iron capacity (million tonnes)	137.2	135,0	135,0	135,0
Coke rate in blast furnaces(kg/t)	514	510	495	485
Other coke consumption (kg/t) (sintering etc....)	84	75	70	65
TOTAL STEEL	598	585	565	550
Coke requirements (million tonnes)				
at a rate of 65%	53,6	51,3	49,2	47,9
70%	-	55,3	53,0	51,6
80%	-	63,2	60,6	59,0
90% utilization of total pig iron capacity	-	71,1	68,1	66,3

Graph.6 Coke Requirements of the Steel Industry (Eur-9)
(Schematic)

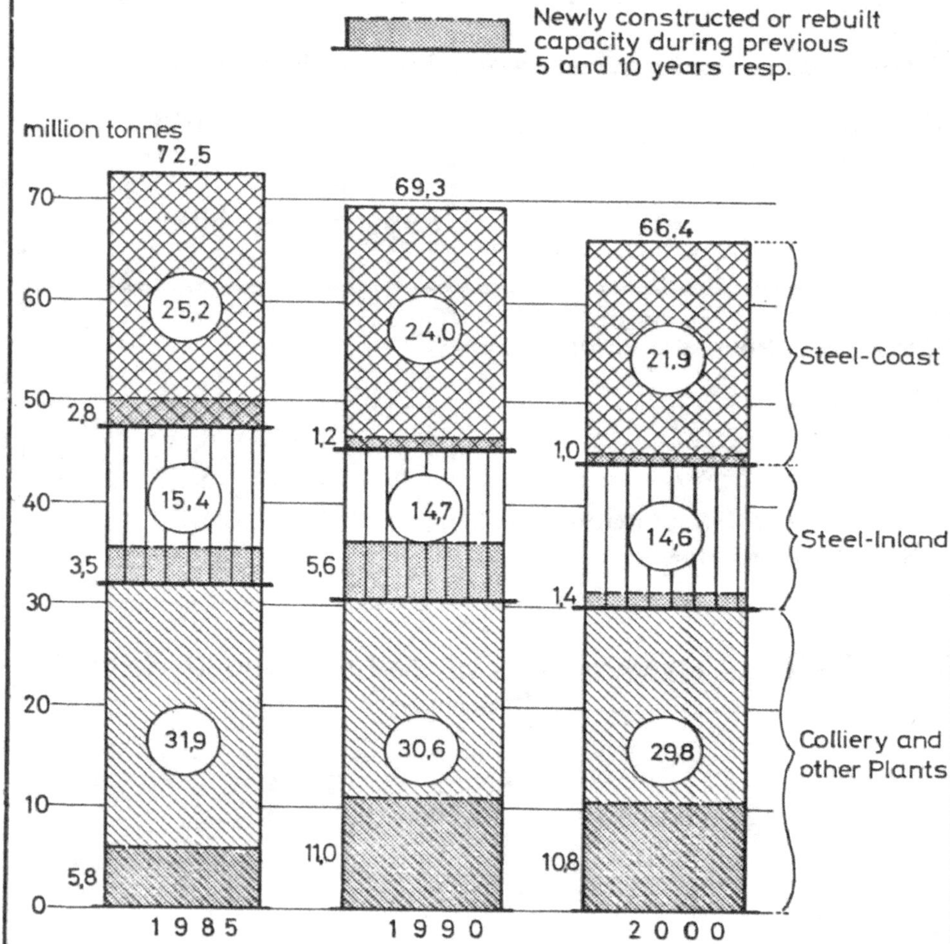

Graph.7 Development of Coke-Oven Capacity (Eur-9)

million tonnes

	1985	1990	2000
Coke-oven capacity	72,6	69,3	66,4
Coke consumption other than steel	(11,3)	(10,5)	(9,5)
Capacity available for steel	61,3	58,8	56,9
Coke requirements at a rate of 70%	55,3	53,0	51,6
80% utilization of total pig iron capacity	63,2	60,6	59,0

Graph.8 Balance of Coke Demand and Supply (Eur-9)

COKE AND COAL IN THE CHANGING TECHNOLOGY OF THE BLAST FURNACE

G. Sironi

Director, Centro Sperimentale Metallurgico SpA

Summary

The steadily rising cost of hydrocarbons in relation to coal has resulted in a rapid shift towards coke- and coal-fuelled blast furnace operations. This trend has been associated with the most recent developments in blast furnace technology, i.e. the revolving chute charging system and the use of blast temperatures over 1200°C. The role of coke characteristics is discussed in the context of the need for continuous refinement of blast furnace technology. The importance of the following points is stressed: use of the size distribution of the coke in order to form charge layers varying in permeability along the radius of the blast furnace and to influence the shape of the melting zone; coke reactivity and coke size degradation as it descends down the blast furnace; knowledge of the links between the parent coals, the coking conditions and the coke structure. The characteristics of formed coke can be controlled during the production process and can, therefore, be adapted to the utilization process. As regards the replacement of fuel oil injection, the problems connected with rapid coal combustion in the tuyeres are examined. An examination is also made of the possibilities offered by the new coal-oil-mixtures (COM) and coal-water-mixtures (CWM), which are still a field for research. Coal-water-mixtures would also make it possible greatly to simplify injection equipment for blast furnaces.

The fact that hydrocarbons (fuel oil and natural gas) are now more costly than coal, combined with predictions that this price differential will be maintained in the future in spite of increased coal demand for energy production, means that hydrocarbons are rapidly being eliminated from blast furnace operations. Fuel oil injection in blast furnaces used to have a number of economic advantages, and it facilitated certain significant improvements in the running of blast furnaces. In general, one kg of fuel oil was able to replace 1.3 kg of coke. Furthermore, fuel oil injection made it possible:

(1) to use high blast temperatures, to control the combustion temperature (by utilizing the heat consumption associated with oil cracking), and to reduce the overall heat cost of the process;

(2) to increase the productivity of the blast furnace by means of an overall reduction in the specific fuel consumption and thus in the blast and in the gas which correspondingly passes through the blast furnace. It is clear that the following factors have made an important contribution here:

a) the improvement in techniques for preparing the ferruginous charge (sinter and pellets) and the coke,

b) the increase in the operating pressure of blast furnaces, and,

c) improvements in the equipment - both in general, and in the systems for charging, controlling and regulating blast furnaces in particular.

It is within this framework of problems and changes that I shall examine the role of coke and coal in the return to operating blast furnaces entirely on coal.

COKE IN THE BLAST FURNACE - CHARGING

The significant advantage of the most recent blast furnace charging systems (which have moveable plates and revolving chutes) is that, by offering a wide range of methods of changing the distribution of the charging materials, they allow the permeability and the thermal capacity of the burden to be adjusted along the radius. In distributing the charging materials, the objectives are as follows:

- to achieve a predetermined top gas temperature, velocity and range profile, in relation to the diameter of the blast furnace;
- to influence the temperature, reduction, profile and level (in relation to the tuyeres) of the melting zone of the burden;
- to protect the blast furnace walls by reducing the heat load by means of a decrease in the gas flow in the zones nearest the walls and by limiting the extent of the area of contact between the melting materials and the walls of the bosh.

In order to achieve these objectives, adjustments can be made to the relative distribution of the ferruginous materials and of the coke. In this way, layers are obtained so that, along the radius of the blast furnace, there are variations in the ferruginous materials/coke ratio, whilst within the same ratio there are variations in the size of the ferruginous materials and coke, which affect the local permeability. The charge distribution, the melting zone and the temperature profile of the blast furnace are shown in Figure 1.

The coke, therefore, plays a key role in determining the physical structure of the burden. In this case the relevant characteristics of the coke are the following: the coke size, and, in particular, the mean size; variations in the coke size and in the mean size; fines content, e.g. - 20 mm; moisture content; mechanical strength (impact resistance and abrasion resistance at ambient temperature, measured by means of a wide variety of tests - I_{20}, M_{40}, M_{10}, I_{10} or DI^{150}_{15}. Table 1 shows, by way of example, the characteristics of the coke charged at Italsider's Taranto no. 5 blast furnace, during part of 1980 (13).

The connections between the mechanical characteristics of cold coke and its production conditions (i.e. blends and coking methods) on the one hand and its behaviour in the blast furnace on the other have already been discussed in ample detail.

Figure 2 shows two examples of charge distribution in which the ratio between ferruginous materials and coke varies along the blast furnace radius: it is high by the walls and low in the centre, in order to encourage concentration of the reactions at the centre of the furnace, to protect the walls and to cause the formation of a softening zone in the form of an acute Λ, which stabilizes the permeability of the furnace at a good overall level and results in high productivity (4). Distributions of this sort are not ideal for obtaining the highest utilization rate of the reducing gases (CO and H_2). In fact, the specific volume of reducing gases and the temperatures are lower, over a good part of the height of the shaft, in the annular peripheral zones where the concentration of ferruginous materials is higher.

Table 1 Coke charged at Italsider's Taranto no. 5 blast furnace during part of 1980. Average production of pig iron 8700 t/24h; coke rate 425 kg/t; specific fuel oil consumption 53 kg/t.

	Variable				Variable		
Chemical analysis	Fixed carbon	89.13	0.52	Other indices	IRSID drum test		
	V.M.	0.48	0.12		I_{20}	82.8	0.3
	Ash	10.38	0.46		I_{10}	15.7	0.4
	S	0.71	0.02		JIS drum test		
					D^{150}_{15}	84.8	0.7
Size analysis	- 20 mm	0.6	0.2				
	20-25 mm	0.5	0.3		ITALSIDER reactivity		
	25-30 mm	2.6	0.6				
	30-35 mm	7.4	0.9		Total weight loss	16.8	2.0
	35-40 mm	6.0	0.9		+ 6.3 mm	90.2	1.1
	40-50 mm	31.6	2.4		- 10 mm	15.1	2.7
	50-60 mm	27.7	1.5		- 1 mm	8.7	0.9
	60-75 mm	18.6	2.2		- 0.5 mm	7.8	0.8
	+ 75 mm	5.0	1.4				
	Mean size	51.9	1.2		NSC reactivity		
	- 35 mm	11.1	1.4				
	- 40 mm	17.1	2.1		Total weight loss	10.0	0.6
	- 40/60 mm	0.75	0.17		D^{150}_{15}	78.2	1.0
	- 40/75 mm	4.10	1.74				

The graph shown in Figure 3 is taken from the study by M. Giuli et al (1), and shows changes in the pressure drop across a one metre thick layer of sinter or coke as its size analysis varies.

More recently, the tendency has been to even out sharp variations in the ferruginous materials/coke ratio along the radius by means of size segregation of the coke, which can be achieved using a revolving chute. In this way, it is possible to place the smaller-sized fraction of the coke in the outer part of the cross section of the blast furnace, which leads to an improvement in the utilization of the reducing gases. In fact, data collected by the CSM (Metallurgical Test Centre) during the charging of a blast furnace fitted with a charging machine with a revolving chute suggests that the coke is segregated in the trajectory from the chute, the larger-sized coke falling nearer the wall (Figure 4). However, these larger sizes roll towards the centre of the furnace, so that the size distribution of the coke is as shown in Figure 5, with the larger-sized coke situated towards the centre.

COKE IN THE STACK OF THE BLAST FURNACE

The burden descends down the stack of the blast furnace at a rate proportional to the output of the furnace, i.e. proportional to the volume of blast blown per unit of time. The coke is subject to mechanical stresses due to the weight of the material above it, the strains between the burden and the furnace walls, the upward pressure of the ascending gases, and, finally, increases in temperature, which can be very rapid in certain zones (Figure 1).

Above a certain temperature level, the chemical action of the water vapour and the CO_2 commences. The kinetics of the chemical attack on the coke are a function of its "reactivity". This combination of mechanical and chemical actions starts off, in the lower part of the stack, the process of coke decomposition which develops to its fullest extent in the bosh (31), (33).

The mechanical, thermal and chemical process of the coke's descent down the blast furnace, for all that I have just said, varies according to the location of the material on the radius of the furnace. Assuming that the rate at which the burden descends is between 0.05 and 0.15 m/min, then the coke at the centre will reach 900°C (the temperature at which it normally starts to be actively attacked by the CO_2) in less than one hour, whilst the coke situated two-thirds of the way along the radius from the centre of the furnace will reach this temperature after two hours.

COKE IN THE SOFTENING ZONE DOWN TO THE TUYERES

Once the coke reaches the melting zone, apart from the passage of the gases, it should be possible to drain off the slag, the metal and the iron oxides in liquid phase (Figure 1). The melting of the ferruginous materials causes a conical pile of coke to form at the centre of the bosh of the furnace. In this zone, the reduction of the iron oxides is completed, the CO_2 produced is transformed into CO, consuming the C provided by the coke (solution loss), the C solution passes into the metal, the respective oxides of Mn and Si are reduced and desulphurization takes place. In the hearth, the C, Mn, Si and S contents of the pig iron assume the final values specified in the casting. Lastly, at the tuyeres, the decomposition of the coke by combustion and gasification is accompanied by the mechanical action of the jet of air, which has a velocity of over 200 m/s. The most recent research confirms that the combustion zone at the tuyeres is surrounded by small-sized coke, precisely because of the chemical and mechanical decomposition caused by the blast (2), (11).

COKE REACTIVITY

Beyond the 900 and 1000°C isotherms, the coke is subjected to chemical action, the main features of which have already been described, and which may be summarized by the following equations:

$$FeO + CO = Fe + CO_2 \tag{1}$$

$$CO_2 + C = 2CO \tag{2}$$

$$H_2O + C = H_2 + CO \tag{3}$$

As from the temperature levels indicated above, reaction (2) is the dominant one. The kinetics of this reaction have been studied by various research workers, with markedly similar results. The rate of the reaction may be expressed by Langmuir's absorption isotherm:

$$\bar{r} = \frac{k_1 \, p_{CO_2}}{1 + k_2 \, p_{CO} + k_3 \, p_{CO_2}} (kg \; kg^{-1} \; s^{-1}) \tag{4}$$

where k_1, k_2, and k_3 may be expressed in accordance with Arrhenius' relationship (5):

$$k_i = A_i \exp(-E_i/RT) \; (kg \; kg^{-1} \; s^{-1} \; kPa^{-n}) \tag{5}$$

The values of A_i and E_i are deduced by obtaining test data on the kinetics of the attack on the coke, using mixtures of CO_2/N_2 and $CO/CO_2/N_2$ at various temperature levels. Examples of the values of A_i and E_i in accordance with equation (5), for various types of fuel and in particular coke and coal, are given by L.G. Austin and L.G. Walker (5), by M. Hatano et al. (6) and by F.D. Skinner and L.D. Smoot (7). Ultimately, the reactivity of a coke can only be determined experimentally, in doing which the k_i and $\bar{r}$ values can be used with certainty for the materials and fields tested.

M. Hatano et al. (6) have also worked out a relationship which expresses the rate of the reaction R due to the attack of the CO_2, taking account of the coke size Dp (mm) in accordance with the equation suggested by Taguchie Okabe (8) and taking account of the type of coke, using the percentage of the solution loss, SL, as compared with that for a reference coke based on test data. According to Hatano, Miyozaki and Jwanage, R may be expressed in a relationship of this type:

$$R = \frac{k_1\ p_{CO_2}}{1 + k_2\ p_{CO} + k_3\ p_{CO_2}}\ f\ (Dp)\ k\ (SL)\ (kg\ kg^{-1}\ s^{-1}) \qquad (6)$$

It is evident that this is an example of a theoretical/ empirical relationship expressing the reaction rate R, and that it is highly dependent on the test data. Research work has now returned to models for predicting R, in which the physical and petrographic characteristics of the coke can be directly used and expressed (36), (42). As a rule, it may be said that an increase in the reactivity of the coke (to CO_2 and to H_2O) involves a corresponding increase in its size degradation as it descends down the furnace; this is due to the weakening of the structure of the coke itself over 1000^oC because of the removal of C in reactions (2) and (3).

There is a large body of test data - including particularly important data obtained in Japan from cooled furnaces in which the entire stock column was analysed (without, however, being able to isolate the effect of the cooling method on such results) - which indicates that the average coke size is practically halved (6) as it descends from the throat of the blast furnace to the bosh. Finally, the permeability of the blast furnace is considerably affected by the degradation associated with the reactivity of the coke, as has been demonstrated by the operation of Italsider's Taranto no. 5 blast furnace, which has a 14 m hearth (10). The reaction is shown in Figure 6.

ALKALIS AND COKE REACTIVITY

In addition to the intrinsic reactivity of the coke, there is the effect of alkalis (either free or combined), which - especially if they are free - make the coke more subject to degradation. Experimental work in laboratories and research on Japanese blast furnaces (3), (4) have indicated that in the stack-bosh zone of a blast furnace, there may be accumulations of alkalis five times greater than the average level in the burden. This concentration of alkalis causes a significant increase in the kinetics of the attack of the CO_2 on the coke, which starts to be more important at lower temperatures, i.e. at 700 to 800°C (instead of at 900 to 1000°C). A further factor in the decomposition of the coke would appear to be the compounds formed between the alkalis and the carbon (31). It is, therefore, said that with the exception of the alkalis contained in the coke itself, the alkali problem generally affects blast furnace operations more than coke oven operations. In fact, however, the alkali content in blast furnaces can be limited by two basic measures (9):

(1) controlled limitation of the specific quantity of alkalis in the blast furnace by appropriate selection of raw materials;
(2) avoiding or eliminating accumulations of alkalis in the blast furnace, either by maintaining the lowest possible $\frac{CaO + MgO}{SiO_2}$ value in the slag within the limits of other technological and metallurgical requirements, or by making periodic and programmed reductions in the basicity, so as to assist the elimination of alkalis together with the slag itself (9).

REACTIVITY AND COKE CONSUMPTION

Within certain limits, the reaction $CO_2 + C = 2CO$ is

essential for the completion of the iron oxide reduction process. The amount of carbon consumed in direct reduction is determined by the thermal and chemical conditions in which the process is carried out (in other words, the characteristics of the fuels used, the blast temperature and the characteristics of the charge). It is, in fact, possible to calculate a good estimate, for each set of a blast furnace's operating and charging conditions, of the amount of direct reduction represented by the summation of reactions (1) and (2), in order to keep the fuel consumption to a minimum. Figure 7 shows the coke consumption of a blast furnace as a function of the percentage of direct reduction, i.e. the carbon consumed in reactions (1) and (2). In the case of coke charged in a large blast furnace, despite the good average level of its technical indices, it could be demonstrated that variations in reactivity had an adverse effect on the utilization rate of the reducing gas η_{co} (correlated with the direct reduction). Variations in reactivity led to variations of up to a factor of 2 in the η_{co} value, corresponding to variations in the coke rate of the order of 6 to 10 kg/t pig iron (13). In general, if all the other conditions are kept constant, the coke should have chemico-physical characteristics which allow the blast furnace to be operated at or around the minimum fuel consumption identified in Figure 7, and it should maintain its size with as little variation as possible until combustion in the tuyere zone. As the reactivity of the coke is increased, the operating level of the blast furnace shifts to the left in Figure 7, with a resultant rise in the coke rate.

REACTIVITY AND STRUCTURAL AND PHYSICAL CHARACTERISTICS OF COKE AT HIGH TEMPERATURES

Various experiments have now confirmed the negative effect of an increase in coke reactivity (as measured in laboratory tests) on the coke rate and on the production of blast furnaces. On the other hand, the relationship between the structure of a coke and its behaviour in the blast furnace is still a field for research. In particular, great importance should be attached - also as regards technological progress in coking

- to the acquisition of more complete knowledge concerning the links between the physical structure of coke (type and distribution of the pores, thickness of the pore walls) and its petrography (in particular carbon petrography) on the one hand and the behaviour of the coke when subjected to chemical attack by gases over 1000°C (12), (31) on the other.

We already have information on the influence of the petrographic characteristics of the parent coals - in particular their vitrinite reflectance - on the porosity and thickness of the coke pore walls, as reported in the work of Benedict and Thompson (12). The porosity decreases and the pore wall thickness increases in direct correlation with the reflectance. The lower the porosity and the greater the thickness of the pore walls, the lower the reactivity of the coke and the greater its mechanical strength, assuming the same thermal and chemical treatment (Figure 8). Therefore, limited porosity and increased thickness of the walls, and increased reflectance of the coke which forms the walls, are the physical characteristics which enhance the coke's mechanical strength and the maintenance of its size as it descends down the blast furnace. In Figure 8 the connection between the reactivity of the coke and the petrographic characteristics of the parent coals can clearly be seen.

The reactivity of the coke can be influenced by increasing the density of the charge, which can be done either by preheating the charge (with stamp charging or partial briquetting of the charge) or by increasing the coking temperature and allowing a suitable soaking time (32). The research work in progress, including some being supported by the EEC, concerning the nature of the coking process and in particular the mesophase (14), (15) and the factors which together determine the porosity and the thickness of the coke walls, may be of great assistance in improving the characteristics of hot coke and coking technology in general.

FORMED COKE

The development in future years of processes for producing formed coke will be linked to a number of factors:

a) the development of the steel industry and its ability to make radical - if gradual - changes in its equipment (in particular, the transition from conventional coking plants to new processes for making formed coke);

b) the pressure of demand on the coal market and in particular the continuing price advantage of non-coking coals acceptable to the steel industry as compared with true coking coals;

c) a probable concentration on a few of the processes for producing formed coke, out of the fairly substantial number of processes potentially available.

The processes for producing formed coke can be divided into two groups:

(1) those with hot forming, which make use of the plastic properties of one or more coals; these include the Consolidation (USA), Ukhin-Giprokoks (USSR), BFL and ANCIT (Federal Germany) processes;

(2) those with cold forming, with a binder: Nippon Steel (16), DKS (Japan), FMC (USA), HNBC (France) and the new CSM process (Italy) (17). In these processes coal or char briquettes are produced with the help of a binder, and various means of carbonizing the briquettes are then employed.

So much for the details of the actual production processes. Although blast furnace tests over a significant period of time have only been carried out on coke produced by a limited number of these processes, it has already been possible to deduce that some types of formed coke can be used to replace equal quantities of traditional coke, even where the characteristics (porosity, density and reactivity) of the formed coke differ from those of the traditional coke (18 - 20).

Referring in particular to the tests done with ANCIT, FMC and HNBC, we can identify not only the problems formed coke may bring, but also the prospects it offers for improvements in blast furnace operations:

(A) Because of the shape of the briquettes, a stack of formed coke may have a lower voids index than sized coke, resulting also in low permeability in the zone of contact with the sinter burden. Thus even if only part of the traditional coke is replaced by formed coke, there may be a certain drop in the permeability of the bosh, leading to some difficulty in keeping up productivity levels when operating with coke alone (43).

(B) However, experimental work has demonstrated that it is possible to produce formed coke with a lower porosity and a higher density than conventional coke, so that it keeps its shape right down to the tuyeres.

Formed coke production processes, therefore, seem to offer particular means of controlling the ultimate properties of the coke throughout the blast furnace process. This possibility seems to be especially marked in the processes involving cold briquetting of char (as in the CSM process), where, by adjusting the size distribution and production conditions of the char, the type of binder, the briquetting pressure and lastly the carbonization conditions, it is possible to fundamentally alter the properties of the formed coke.

The fact that, compared with conventional coke, a burden made up of formed coke briquettes of one type only has a lower voids index, can be compensated for in an industrial situation by using briquettes of various shapes and sizes mixed together, or by employing special charging techniques (43), (46), (47).

COKE-ONLY BLAST FURNACE OPERATION

The move away from injecting hydrocarbons in tuyeres has necessitated a certain amount of readjustment in the operation of blast furnaces, which in most cases is based on the expectation that 50 to 80 kg of fuel oil is injected per tonne of pig iron produced. In this context, it has been considered necessary to re-examine the use of the high temperature blast (1200 to 1300°C) produced in more recent hot blast stoves, so as not to exceed the adiabatic temperatures (2200 to 2300°C) associated with operations employing fuel oil injection.

The return to coke-only operation has often meant a less desirable return to higher coke rates and a tendency for blast furnace production to be reduced as the volume of gas per tonne of pig iron rises, especially when the blast temperature is lowered in order to limit the combustion temperature. Experience in Japan has been of an average drop in production of about 5% (2). Coke-only operation with high blast temperatures, that is in more modern blast furnaces, is a relatively recent phenomenon and further progress could be made by:

a) improving the distribution of the charge so as to take account of the new conditions, thus taking advantage of the increase in permeability gained by raising the coke/ ferruginous materials ratio;

b) improving the hot coke characteristics so as to prevent degradation in coke size in the bosh;

c) raising the operating pressure of the furnace (if the plant will allow it);

d) making full use of the blast temperature by experimenting with higher combustion temperatures (2350 to 2400°C), possibly controlled by adding steam to the blast air. The addition of steam allows the H_2 content of the reducing gases to be kept at several percent in the absence of hydrocarbons, thus favouring the kinetics of reduction in the stack.

By means of a suitable combination of some of the measures described above, it has been possible, in the cases of Solmer's no. 2 blast furnace, which operates on coke alone, to obtain very promising results (450 kg coke/t pig iron) and to improve on the operating results achieved with fuel oil injections (49).

INJECTION OF COKING BY-PRODUCTS IN THE TUYERES (COKE OVEN GAS AND TAR)

Injections of coke oven gas or tar in the tuyeres may be envisaged as a means of fully exploiting the preheating capacity of the blast. The injection of coke oven gases necessitates the construction of special equipment fitted with compressors capable of bringing the gas to the pressure required for injecting it into the blast furnace. This equipment is more costly than comparable equipment for injecting fuel oil or natural gas. In Italy, two of Italsider's blast furnaces at Trieste are equipped for coke oven gas injection (22). In the no. 2 blast furnace (April 1981), the gas injection rate was 79 Nm^3/t pig iron, and the coke rate 458 kg/t (the blast temperature was 1000°C, and foundry pig iron with 1.3% Si and 1.1% Mn was produced). Nevertheless, the injection of coke oven gas is not widespread, either because it is used far more as a reducing gas in blast furnace tuyeres or because at most plants the exploitation of the heat content of the gas is a fully integrated part of the operational cycle.

Much the same applies to tar, which can be injected into blast furnaces by means of equipment similar to that used for fuel oil. At present, there are various blast furnaces operating with tar injection; one example is Italsider's Taranto no. 3 blast furnace (hearth diameter 10.3 m), which has been operating since mid-1980 with an injection rate of about 50 kg/t iron, and with a replacement ratio estimated at about 1.1 kg coke/kg tar at a blast temperature of 1150°C. However, here too the use of tar as a chemical feedstock seems to be given preference; tar utilization in blast furnaces is, therefore, a function of comparisons of costs and proceeds between tar, fuel oil and coke.

COAL INJECTION IN BLAST FURNACES

The injection of coal in blast furnaces is the technique which is currently being studied with the greatest interest; it is seen as a means of eliminating hydrocarbons from blast furnaces and of fully exploiting the high blast temperatures possible with modern stoves; it may possibly also be linked with enrichment of the blast with oxygen. Coal injection was tried out in Community countries back in the 1960's (23), (24), but was not developed any further in view of the competitive price level of hydrocarbons and the simplicity and low cost of equipment for injecting fuel oil and natural gas.

Further tests with coal injection have now been carried out in the USA (ARMCO) (25), the People's Republic of China (SICS) (26) and the USSR (27) and have commenced in Europe (ARBED, Luxembourg). The data available confirms the estimates already made on the basis of calculations and of the first test runs. The coal/coke replacement ratio achievable varies, according to the type of coal and of blast furnace, from 0.9 to 1.1. The quantity of coal injected reached 150 to 270 kg/t pig iron during test periods, with up to 25% enrichment of the blast with oxygen (SISC). In one Chinese blast furnace, the most recent (1979) average is around 150 kg/t pig iron, and the enrichment of the blast with oxygen over 23%. At ARMCO, the carbon injection rate is some 80 kg/t pig iron. Lastly, it should be pointed out that in Luxembourg, ARBED are operating new equipment for injecting dried lignite or coal.

The industrial data now available is in some cases backed up by years of experience, but it relates to small to medium blast furnaces whose operational characteristics date back to the 1960's. For this reason, it is unlikely that the body of operating experience now available will be either sufficient or immediately transferable to larger blast furnaces with hearth diameters of 10 m or more. Further research work and industrial trials will, therefore, be necessary.

REPLACEMENT RATIO AND AMOUNT OF COAL INJECTABLE

In the case of a blast furnace operating at a high blast temperature (1000°C), with fuel oil injection, it is possible to maintain the same adiabatic combustion temperature by replacing the fuel oil with a corresponding or greater quantity of coal, which is estimated on the basis of a heat balance of the combustion zone.

For this purpose, it is useful to compare the heat of decomposition of fuel oil, tar and coal as a function of the percentage of H_2 (Figure 9). For coals with a volatile matter content around 30 to 35%, the heat of decomposition is around 850 KJ/kg, i.e. half the average for fuel oils; for anthracitic coals, the heat of decomposition is some tens of KJ/kg less.

It can be deduced that if the theoretical combustion temperature is kept constant, more coal can be injected in place of fuel oil the lower the volatile matter content of the coal is. Thus, taking account of this and on the basis of the experience of the Chinese (26), anthracitic coal allows higher injection rates, and the same may also apply to char.

COMBUSTION OF THE COAL INJECTED IN THE TUYERES

The combustion of coal in the tuyeres of large modern blast furnaces presents a variety of problems, and the best means of solving these is still a field for research.

Type of coal: low V.M. coals and (always) low-ash coals certainly have the advantage mentioned above of a lower heat of decomposition, but more often than not a wide variety of medium and high V.M. coals are used (e.g. ARMCO and USSR).

Size of coal injected and combustion rate: in the trials conducted (35) in blast furnaces in France (23) and the United Kingdom (24), relatively coarse coal sizes have been used; in fact at Chasse (23), the coal was ground to -2 mm, and at Stanton (24), the size was -2.36 mm. At present, however, very fine coal is being injected: ARMCO is using coal with 80% less than 200 mesh (0.075 mm). The SISC's blast furnaces, although no precise information is available (26), must be using coal with 75% less than 0.074 mm and 100% less than 0.5 mm. ARBED (of Luxembourg) are proposing to use coals sized to 90% less than 0.090 mm. Although the use of very fine coal sizes involves substantial grinding costs (capital cost and energy cost), it does ensure a very high combustion rate. According to SISC data, for coal sizes less than 0.075 mm, the combustion time is between 0.01 and 0.03 seconds. Data relating to ARMCO's blast furnace tuyeres indicates that these combustion times are even shorter, virtually 10 times less than those just mentioned, i.e. between 0.001 and 0.005 seconds; this data probably relates to char. It would, therefore, be interesting to analyse the theoretical and experimental work done on the kinetics of combustion of pulverized coal in conditions similar to those existing in blast furnace tuyeres, so as to assess the possibilities for improving this technology and better adapting it to use in blast furnaces.

The study undertaken by R.H. Essenhigh, R. Froberg and J.B. Howard (28) confirmed the hypothesis that the combustion of small coal particles (as in the case of coal injected in blast furnaces) is a sequence of three phenomena (or resistances): diffusion of the O_2 through the boundary film of the particle, adsorption of the O_2 and finally desorption of the product of the reaction. The reaction rates of these individual phenomena, in the order k_o, k_1 and k_2, are interrelated on the basis of a Langmuir isobar adjusted in accordance with the hypotheses derived from the test data. The individual k values which together express the reaction rate are expressed and calculated in accordance with Arrhenus' relationship, as in the case of the kinetics of gasification.

The curves shown in Figure 10 (28) are particularly interesting technologically speaking, as they indicate the order of magnitude of the specific combustion rate R_s and also to what extent it may be increased. In fact, the combustion rate can be speeded up by raising either the temperature T or the oxygen concentration in the blast. Thus in the blast furnace process, the use of high blast temperatures and enrichment of the blast with oxygen, as practised by the SISC, are very favourable to rapid coal combustion. The ability to predict the combustion time of a coal particle as a function of its diameter is useful for estimating whether the particle will be consumed within the oxidation zone, in front of the tuyeres or in the "raceway" space. According to Peters (48), the degassing time τ(sec) of a coal particle of diameter d (mm) and V.M. content M (%), in a gaseous medium at temperature t (^{o}C), is given - according to (26) - by

$$\tau = \frac{Md^{0.26}}{0{,}03(t-330)} \qquad \text{(sec)}$$

from which it will be seen that the higher the V.M. content, the slower the combustion process, where the combustion process means the heating time, the degassing time and the char combustion time. The combustion time for char τ_{ch} (sec) has been estimated by Field (29), on the basis of the following relationship, which is given in (30):

$$\tau_{ch} = \rho_{c,\ ch}\ R_g d^2_{ch}/(96\ \phi\ Dp_{O_2}) \qquad \text{(sec)}$$

in which $\rho_{c,\ ch}$ represents the carbon content of the char (g/cm^3); R_g the gas constant (82.06 atm cm^3/g mol ^oK); d_{ch} the diameter of the char particle (cm), ϕ a factor relating to the char combustion process; D the oxygen-nitrogen molecule diffusion speed in cm^2/sec; and p_{O_2} the partial oxygen pressure (atm). The same char combustion time can be calculated on the basis of the following relationship, taken from Hedden:

$$\tau_{ch} = \rho_{c,\ ch}\ d_{ch}/(72\ \beta C_{O_2}) \qquad \text{(sec)}$$

in which β is the diffusion coefficient (cm/sec) and C_{O_2} is the O_2 concentration in the gaseous atmosphere (mol/cm^3).

INJECTION OF COAL IN A MIXTURE WITH OIL AND/OR WATER (COM OR GWM)

The preparation and combustion of mixtures (37), (38) of fuel oil and coal (coal-oil-mixtures) has attracted great attention in recent years because of the potential this technique offers for replacing 40 to 50% of the fuel oil burnt in the large boilers of thermal power stations. In fact, it is also possible to use stable and unstable, pumpable and injectable mixtures with a coal content of up to 50% in blast furnaces (39). Higher coal proportions lead to very sharp rises in the viscosity of the COM.

The mixtures currently being produced and tested contain between 40 and 50% of ground coal, with 80 to 90% less than 200 mesh (0.075 mm); if an improvement in the stability of the mixture is desired, this can be achieved by grinding the coal even more finely (less than 300 mesh) or by adding additives. The preparation of a COM necessitates a high degree of grinding, dry and in a protective atmosphere, or with the coal already mixed with the oil. For a COM to be transported, it must be kept at a temperature suitable for the fuel oil used (80 to 90°C), specially designed pumps and circuits are necessary, and the characteristics of the materials used have to be studied and tested accordingly. Up to now, COM's have only been used in blast furnaces on a test basis, in a few tuyeres only and for limited periods (37), (39).

Assuming, as may be deduced from the first trials, that the injection of a COM with a coal content of 50% at a rate of 50 to 70 kg/t does not create any particular problems in the blast furnace process, this COM is likely to give a coke replacement ratio somewhere between that of coal and fuel oil, depending mainly on the type of coal. If the combustion temperature at the tuyeres is constant, this replacement ratio will be in the region of 1.1 to 1.2 kg coke/kg COM.

To sum up, in COM's coal could potentially replace about half of the fuel oil which can be injected in blast furnaces; with mixtures of coal, oil and water, up to two-thirds of the fuel oil might be substituted for (40). Given that the preparation and transport equipment normally associated with fuel oil injection cannot be used for COM's, the prospects for such mixtures are still uncertain, even if it proves possible, in developing COM preparation and combustion technology, to make use of the important research and development work being done in connection with combined heat and power plants. A technique which might be especially applicable to blast furnaces is the coal-water-mixture (or coal-water slurry). Reports published (41) and work carried out by the CSM suggest that coal-water slurries can be produced with coal contents of up to 70 to 75%, with viscosities making them suitable for pumping. The preparation of such mixtures is still being researched into, and it is dependent on the use of suitable additives. There is real potential for using these coal-water-mixtures in blast furnaces, where they would allow fuel oil injection to be eliminated completely, and coal injection would mean full utilization of the heating capacity of the most modern hot blast stoves.

CONCLUSIONS

Important developments in blast furnace technology in recent years have been the move away from hydrocarbon injection in tuyeres, the use of revolving chute charging systems, and more sophisticated criteria for blast furnace operations by determining and controlling the melting zone.

Coal and coke already play an absolutely crucial part in the production of pig iron. The characteristics of the coke must, therefore, meet the following requirements:

- The size distribution and the mechanical strength of coke, when cold and when rapidly heated, will need to be compatible with the new charging techniques, which allow the fluid-dynamic and thermal topography of the blast furnace to be better controlled than in the past.
- Resistance to attack by CO_2 and H_2O should be improved, i.e. the reactivity of the coke should be limited, so as to ensure that degradation is kept to the minimum possible as the coke descends to the tuyeres.
- Increased knowledge of the basic factors influencing coke reactivity is desirable, so that the physical structure of the coke itself (porosity, pore wall thickness and petrographic structure) can be modified by suitably adjusting coal input blends and coking conditions.
- Formed coke production processes open up the possibility of using non-coking coals and also make possible new means of controlling the various indices of the coke formed.
- Coal injection is the most attractive technology for replacing hydrocarbons. The experience currently being obtained (by ARMCO, SISC, DONETS and ARBED) will need to be adapted to larger blast furnaces. Further refinement of the coal injection process itself will also be necessary (choice of coal, fineness of grinding, enrichment of the blast with oxygen). It will be possible in the short term to inject 100 kg coal/t pig iron, and maybe more, in normal operation; in this way, 20 to 22% of the coke charge would be replaced.
- Another interesting technique which could be applied in the near future is the injection of coal-oil-mixtures (COM's), even if at the present state of the art injection would be limited to 30 to 50 kg of coal forming 50% of a COM.
- Coal-water-mixtures appear to be potentially more promising

for injection in blast furnaces, especially if it proves possible, in industrial use, to raise their coal content above the present 70%. This technique, which would allow the injection equipment to be considerably simplified, requires a high temperature blast and, possibly, enrichment of the blast with oxygen.

BIBLIOGRAPHY

1) M. Giuli, A. Piccioni, M. Pinti-Boll. Tecnico Finsider, no. 385 March-April 1980, p. 120-132.

2) N. Nakamura, Y. Togino, M. Takeoka - Coal, Coke and the Blast Furnace - The Metals Society, 1977, p. 1-18.

3) M. Hatano, T. Miyazaki, Y. Jwanaga - Trans. ISIJ Vol. 20 no. 9, 1980, p. 592-598.

4) M. Hatano, T. Miyazaki, Y. Jwanaga - Trans. ISIJ, Vol. 20 no. 11, 1980, p. 744-752.

5) L.G. Austin, L.G. Walker - AIChE Journ., Vol. 9, no. 3, may 1963 p. 303-306.

6) M. Hatano, T. Miyazaki, Y. Jwanaja - Trans. ISIJ, Vol. 20, no. 5, p. 292-300.

7) F.D. Skinner, L.D. Douglas Smoot "Heterogeneous Reactions of Char and Carbon "Polverised-Coal Combustion and Gasification" edited by L.D. Smoot and D.T. Pratt - Plenum Press.

8) S. Taguchi, K. Okabe - Kawasaki Steel Technical Report, 2 (1970) p. 358.

9) W.K. Lee, J.E. Holditch, N. Daneliak "Alkali control in the Blast Furnace" BF Conference - HF80 - Vol. 3, Arles (France) 1980.

10) A. Ghiglione, M. Giuli, P. Barnaba, F. D'Aquino "Blast Furnace Conference" Arles (France) June 1980 - HF 80 - Vol. 3.

11) M. Kondoh, Y. Konishi, Y. Morioka, S. Tomita, S. Ashizume - Trans. ISIJ Vol. 20, no. 7, 1980, p. 252.

12) L.G. Benedict, R.R. Thompson - International Journal of Coal Geology, no. 1, 1980, p. 19-34.

13) Italsider-CSM: Gruppo di Lavoro - Rapp. Tecnico no. 4 (1980) Convenzione CEE-7210-AA-401.

14) H. Marsh - AIME IRONMAKING PROCEEDINGS, Vol. 39, Washington DC, p. 266-273.

15) J.J. Friel, S. Mehta, G.D. Mitchell, J.M. Karpinski - AIME Ironmaking Proceedings, Vol. 39, Washington DC, 1980, p. 274-278.

16) T. Okuhara et al. - Nippon Steel Technical Report, no. 15 June 1980, p. 77-93.

17) L. Palumbo, A. Colletta - Bollettino Tecnico Finsider, nov.-dec. 1980, p. 343-350.

18) W. Hermann, R. Schmit - Agglomeration 77, AIME, Vol. 2, p. 970-995.

19) J.A. Cordier, R. Rochas - Agglomeration 77, AIME, Vol. 2, p. 996-1011.

20) P.K. Strangway, M.O. Holowaty - Agglomeration 77, AIME, Vol. 2, p. 1012-1050.

21) S. Yabe - Sumitomo Metal Ind. Ltd. - IISI - 13th Committee on Technology Melbourne, March 1981.

22) A. Ghiglione, M. Giuli, E. Filippone, G. Sironi, R. Sacerdote, L. Tofful, G. Venier - Symposium on Blast Furnace Injection - AIMM, Wollongong (Australia), 1972.

23) P. Chauzat, C. Reveil, J. Cordier, A. Damgè, J.C. Leclère - Revue de Métallurgie, Jan. 1963, p. 39-48.

24) E.M. Summers, L. MacNaughton, J.R. Monson - J. of Iron and Steel Inst. August 1963, p. 666-677.

25) A.K. Garbee - HF 80, Arles (Francia), June 1980, Vol. 3.

26) Guo Yun-zhi, Liu Yun-cai - "Injection de poudre de charbon dans le haut forneau" - Revue de Métallurgie, CIT, Oct. 1980, p. 841-851.

27) L. Yaroshenvskii, V.N. Tereshchenko, G.E. Nekhaev, A.D. Didevich, O.K. Anufriev - Stal 1979, vol. 5, p. 328-331.

28) R.N. Essenhigh, R. Froberg, J.B. Howard - Industrial and Engineering Chemistry, Vol. 57, no. 8, Sept. 1965, p. 33-43.

29) M.A. Field - Combustion & Flame, Vol. 13, June 1969, p. 237-252.

30) R.R. Rajon, C.Y. Wen - AIChE Journ., Vol. 26, no. 4, July 1980, p. 642-655.

31) British Steel Corporation Technical Reports ECSC Agreement, no. 7210-AA-8/801 e 7210-AA/8/802, 1976-1981.

32) K.G. Beck - Stahl u. Eisen 100, no. 3, 1980, 11 February.

33) E. Szurman - Stahl u. Eisen 100, no. 12, 1980, 16 June.

34) H.W. Gudenau, K. Kreibich, K.H. Peters - Stahl u. Eisen 101, no. 3, 1981, 9 February.

35) J. Monson, D. Gathergood - Stahl u. Eisen 101, no. 1, 1981.

36) A. Funghini, P. Fontana, G. De Marchi, B. Canepa, P. Costa - Congres Haut Fourneau, Arles (France) June 1980 - HF 8 - Vol. 2.

37) Proceedings Second International Symposium Coal Oil Mixture Combustion 1979, Vol. 1-2, Pittsburgh Energy Technology Centre, Pittsburgh, PA.

38) Proceedings Third International Symposium Coal-Oil Mixture Combustion April 1-3/1981, Vol. 1-2, Pittsburgh Energy Technology Centre, Pittsburgh, PA.

39) I. Kurashigo et al. - Abstracts 40th Ironmaking Conference, March 29-April 1, 1981, Toronto (Canada), IIS of AIME, p. 123.

40) L. Poetsche - "Liquicoal Demonstration Project", Vol. 1, (38).

41) R.S. Scheffee et al. - "Development and Burning of Coal/ Water Slurries", Vol. 1 (38).

42) J. Szekely, D.A. Aderibidge - Abstracts 40th Ironmaking Conference, March 29-April 1, 1981, Toronto (Canada) - IIS of AIME.

43) R. Limpach - Rapport final Convention CCE no. 6210-AA/5/502, 1979.

44) M. Hatano, O. Kohji, H. Yamaoka, C. Yamagata - Trans. ISIJ, Vol. 20, no. 7, 1980, p. 256.

45) Rapp. Tecn. Convenzione CEE no. 7210-AA/405 Italsider, Sept. 1980.

46) H.J. Schultz, O. Obel - Glückauf, 110, Vol. 13, p. 517-522 (BISI 13043).

47) O. Obel, H.J. Schultz - H1, January 1972, p. 1-11.

48) W. Peters, J. Lehman - "Erdöl und Kohle Erdgas Petrochemie" February 1965, no. 2, pp. 112-118.

49) C. Thirion, N. Jusseau, J. Bouget "Marche tout coke aux hauts forneaux de Solmer", Commission Fonte ATS, Paris 23-24 April 1981.

Charge temperature distribution in the shaft

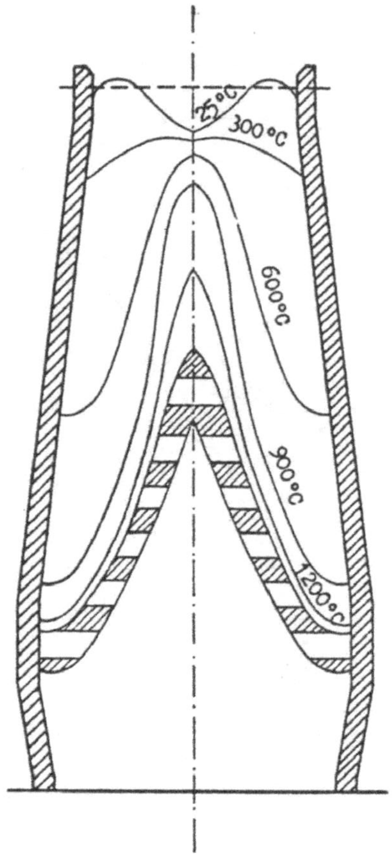

Figure 1 : Section of a blast furnace, showing the isotherms and the melting zone

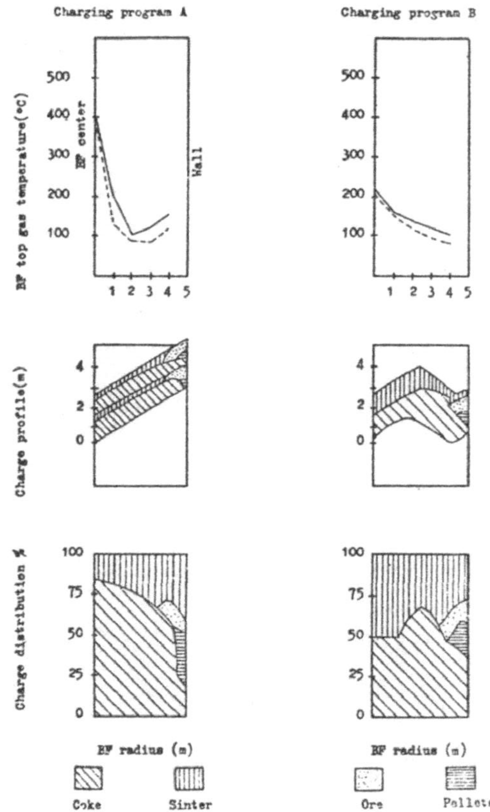

Figure 2 : Two examples of charge distribution where the ferruginous materials/coke ratio varies along the blast furnace radius. Charge distribution by revolving chute.

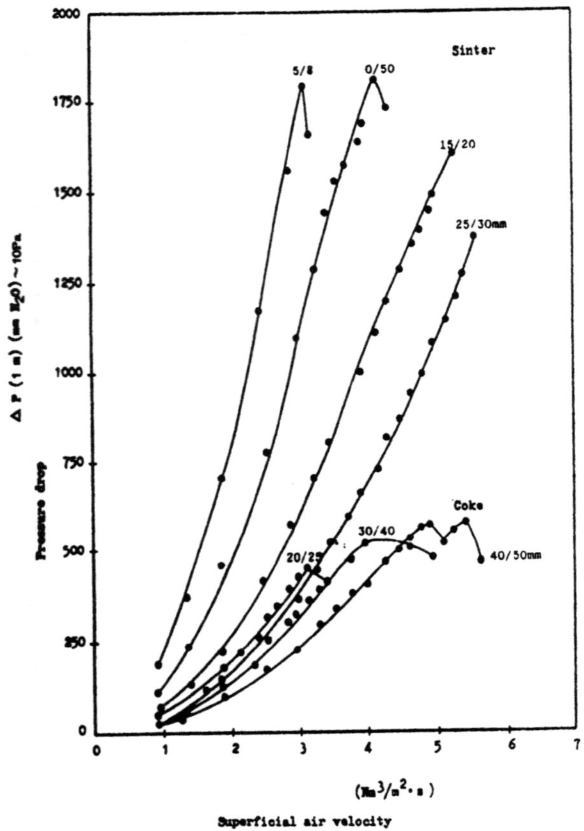

Figure 3 : Pressure drop across a one metre thick layer of sinter or coke with varying ssize distributions

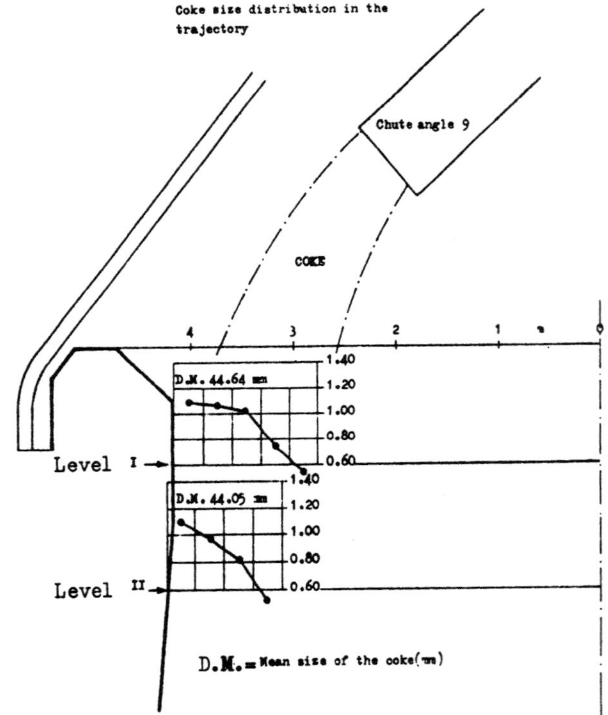

figure 4 : Segregation of the coke as it falls from the revolving charging chute (Giuli, Palchetti - CSM)

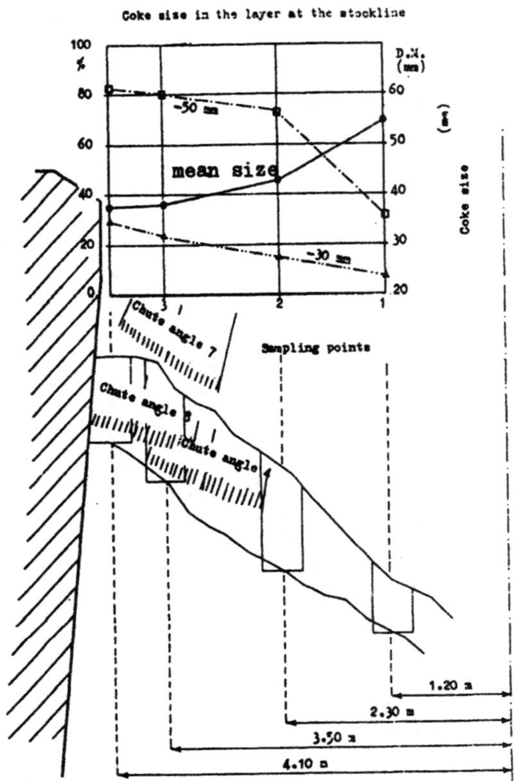

Figure 5 : Segregation of the coke by size after it has settled on the charge below

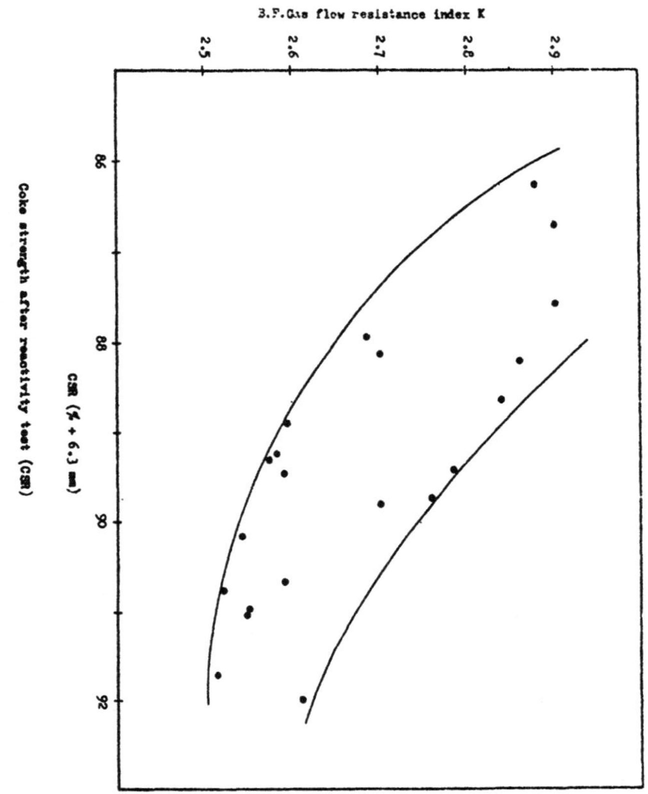

Figure 6 : Relationship between coke strength after reactivity test and the permeability of the blast furnace

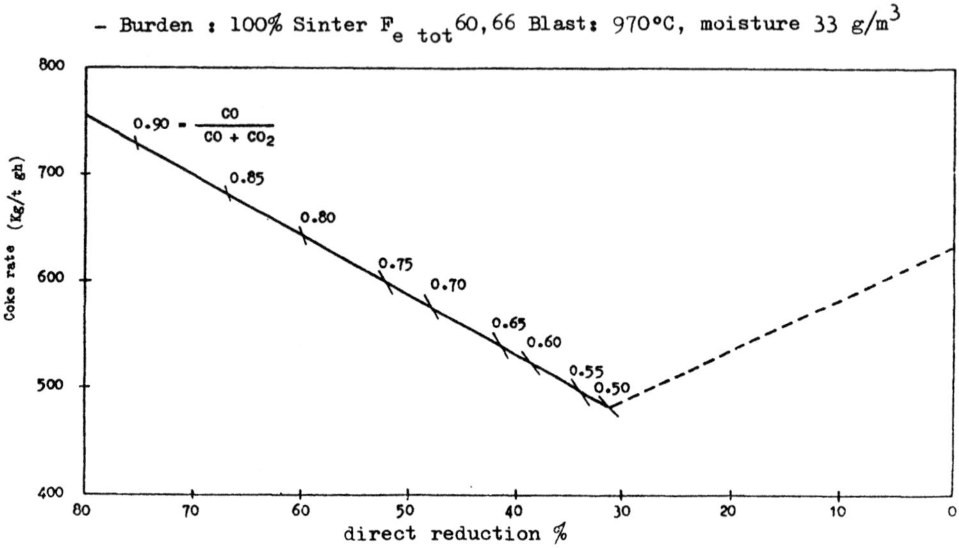

Figure 7. Coke rates in a blast furnace as a function of the direct reduction of the iron oxides ($FeO + CO = CO_2 + C = 2CO$).

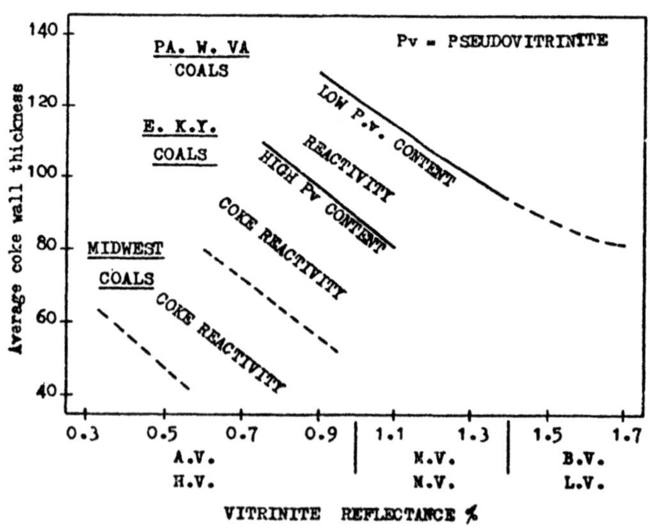

Figure 8. Reactivity of coke as a function of the average pore thickness and of the vitrinite content (reflectance) of the parent coals (Benedict, Thompson, Bethlehem (12)).

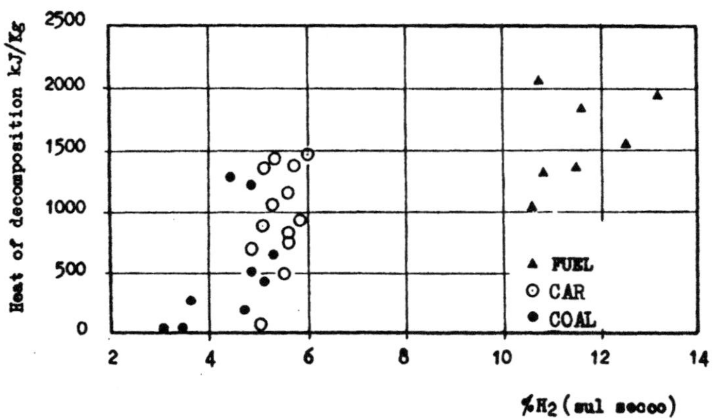

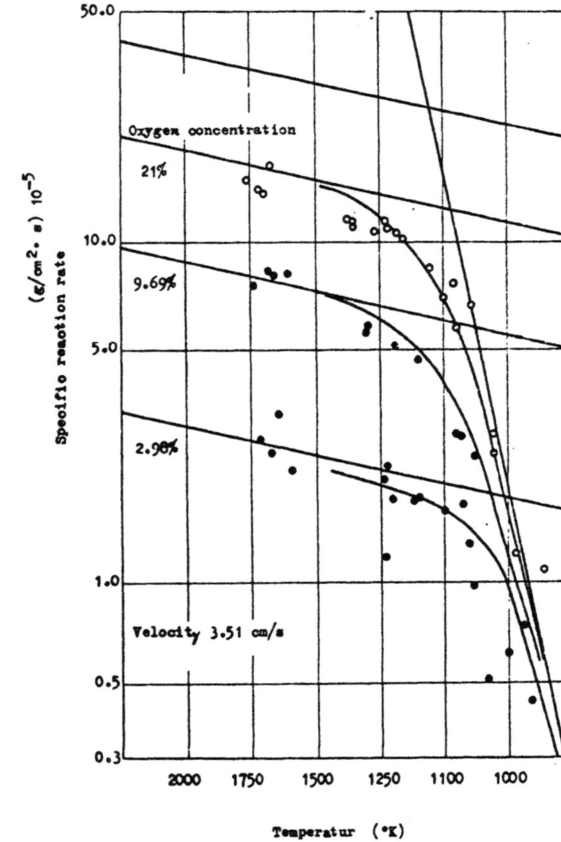

Figure 9 : Heat of decomposition of fuel oil, tar and coals as a function of the percentage of H_2 (dry) (Palchetti - CSM)

Figure 10 : Specific reaction rate of carbon as a function of the ambient temperature and of the oxygen concentration in the blast (Essenhigh, Froberg, Howard (28))

FIRST TECHNICAL SESSION: BASIC RESEARCH

Chairman: G. DELLANOY, Director, Centre d'Etudes
et Recherches des Charbonnages de France

- Studies of the factors controlling the formation and development of the porous structure of coke
  J. W. PATRICK, British Carbonization Research Association

- Coal carbonization: formation, properties and relevance of microstructures in resultant cokes
  H. MARSH, Nothern Carbon Research Laboratories, School of Chemistry, University of Newcastle-upon-Tyne

- Influence of the plastic properties of steam coal by means of bituminous additives
  O. ABEL, Technische Universität Clausthal

- High temperature stability of coke in relation to its bilk permeability
  J. GOLECZKA and J. TUCKER, National Coal Board

- Characterization of coal and coke
  J.M. DUCHENE and G. PRUDHON, Marienau Experimental Station

- Round table "A" Coke and the blast furnace
  Chairman: G. SIRONI, Director, Centro Sperimentale Metallurgico

- Round table "B" Pollution problems
  Chairman: J.P. GRAHAM, Director, British Carbonization Research Association

- Introductory paper
  Emission of tar fog from coke oven doors
  R.F. LITTLEJOHN and M.A. WRIGHT, National Coal Board

FIRST TECHNICAL SESSION

CHAIRMAN'S INTRODUCTION

G. DELLANOY

The first meeting is concerned with basic research on coke. This is of interest primarily to laboratory workers, but I think that it is of even greater importance to coking plant engineers, who have to deal with a very diverse range of problems. Manufacturing parameters have to be correlated with each other and sometimes a choice has to be made between them, in order to meet quality requirements.

We must never forget the customer, although present circumstances might make us do so if we had short memories: increasing stock levels, the steel crisis and the general situation of reduced economic potential, together with the gloomy political climate, mean that the research effort must not falter.

I personally spent more than 20 years in production, was then head of a research laboratory for 13 years, and have now returned to production. Believe me, this hopping over the fence makes you see both points of view. And although I have become a veritable ignoramus in your field I shall pay great attention to the papers which we are about to hear.

STUDIES OF THE FACTORS CONTROLLING THE FORMATION AND DEVELOPMENT OF THE POROUS STRUCTURE OF COKE

J. W. PATRICK

British Carbonization Research Association

Summary

The influence of several carbonization variables on the porous structure of coke has been quantitatively determined using optical microscopy allied to computerized image analysis to determine details of the porous structure. Thus preheating of the charge was shown to result in coke with a greater number of smaller pores which are less widely spaced on average. Increasing the blend vitrinite reflectance generally reduces porosity and pore size whilst the pore-wall size shows little change although this dependence varies with different coals. Additions of low-volatile additives to the blend follows this same general pattern but with certain petroleum cokes and coals of volatile matter less than about 12.5% there is also an increase in the pore-wall size. It has been shown how the data so obtained can be used to formulate blends to produce cokes of specific structural, and hence strength, characteristics.

Studies of the process of development of the porous structure of coke using laboratory scale carbonizations show that although the total dilatation of the coal has some influence on the initial stage of the process, it exerts no dominant influence on the pore structure of the final coke product.

1. INTRODUCTION

The porous structure of coke is of prime importance in controlling its strength (1). Hence in any attempt to either improve the strength of metallurgical coke or to consider the production of coke of specified strength characteristics it is of considerable advantage to have some knowledge of the factors influencing the formation and development of the porous structure of coke. However the development of the porous structure during the carbonization of coal under conditions pertinent to coke oven practice has not been subjected to a comprehensive, detailed quantitative investigation, although some useful guidelines have been reported previously (2, 3). The acquisition of new techniques by means of which the required structural information can be obtained has facilitated to a large extent the attainment of this knowledge.

Thus the purpose of the work to be described was to obtain an appreciation of the relationship between on the one hand, the coal-charge properties and carbonizing conditions and, on the other hand, the development of the porous structure of coke, and to put on a systematic, quantitative basis, the influence of the related coking variables with a view to ultimately controlling and improving the industrially significant physical properties of metallurgical coke.

2. EXPERIMENTAL PROCEDURES USED

Several techniques were used in this work but the main method used in the examination of the structure was based on the use of optical microscopy in the reflected light mode and analysis of the images by a computerized image analyzer, a Quantimet 720, to give values for a series of structural parameters (4). The parameters measured include those assessed by traditional manual methods, namely, volume porosity, pore and pore-wall intercept sizing and a count of the number of pores, average values being obtained from the analysis of 200 fields of view for each sample. With these procedures the standard errors involved in the measurements were of the order of ± 0.5% for the porosity, ± 3.0 and 1.5 μm for the pore and pore-wall sizing respectively and ± 4 for the number of pores per field of 5.95 mm^2. At the low magnification chosen the absolute limits of measurement extend from 5.45 μm to 4 mm linear size.

In addition, measurements were made on each individual pore greater than 0.0149 mm^2 in area, these measurements being of Feret's diameters i.e. caliper diameters, in sixteen directions, the largest and smallest of these diameters being equatable respectively with the maximum and minimum dimensions of a pore.

To obtain a representative sample for the structural examination, the procedure followed for test-oven cokes was as described previously (5). A 25 kg sample was used to provide a bulk sample of 1 cm diameter discs, core-drilled from all coke pieces of adequate size, i.e. about 2 cm, and 50 - 60 of these discs of approximately 1 cm thickness when mounted in epoxy resin and prepared as polished blocks, constituted the sample for the structural examination. Care was taken to ensure that a high degree of polish free from relief was obtained and standard methods of sample division were used throughout to ensure the representative nature of the sample (4). The 1 cm diameter discs also formed the specimens used for the determination of the coke tensile strength by the diametral-compression test (6).

The cokes examined were generally produced either in the BCRA test ovens of capacity ranging from 17 t down to 250 kg or on the laboratory scale in either a single-side-heated-wall oven which provides the facility to study the solid products from carbonization at a range of temperatures between 300°C to 1000°C or a tubular furnace designed to enable a hot zone to be progressively moved over a coal charge (7).

3. INFLUENCE OF PREHEATING OF THE CHARGE

From the examination of a wide range of cokes, it is now clearly established that preheating as practised on the industrial scale leads to the production of coke which, compared with coke from the same blend charged wet, has a greater number of smaller pores which, on average, are less widely spaced (5,8). Although the coke porosity is frequently reduced this is not an invariable consequence of the preheating of the charge. These changes provide the explanation for the well documented improvement in coke quality which generally accrues from preheating (9).

The magnitude of the structural changes varies considerably, however and there are several factors which could be expected to influence or contribute to their magnitude. Thus the method of charging has been shown to be a factor with the reduction in pore size and increase in the number of pores being less pronounced with a pipeline charge than with a preheated-gravity charge, although the magnitude of these changes is generally small in comparison with those due to the preheating of the charge generally (8).

Another obvious factor which would be expected to have a bearing on the magnitude of the structural changes arising from preheating, is the 'rank' of the coal blend charged since preheating generally is more beneficial, in terms of its effect on coke quality, with charges of lower rank coals (10). However the results obtained to-date whilst giving some indications of possible trends give no definite relationships.

Thus the magnitude of the changes in the structural parameters due to preheating is variable but on average the changes are in the order of 22% decrease in pore intercept size, 20% increase in the number of pores and 12% decrease in the pore-wall size.

4. INFLUENCE OF 'RANK' OF BLEND CARBONIZED

For coal blends typical of those used commerically, the general effect of increasing the blend rank, as measured by the vitrinite reflectance, is to diminish the porosity and the pore size whilst there is no systematic trend with pore-wall size. There is also some indication of an increase in the number of pores. An example of this behaviour is shown in Fig. 1 for a series of cokes produced from blends of various proportions of four coals.

It is however also clear that factors other than the rank as measured by the vitrinite reflectance have to be considered as shown by the data in Fig. 2 which refers to another series of cokes from blends of three coals. This shows that the rank dependence varies with some other, at present unidentified, factor associated with the blend composition. Nevertheless the structural values follow in general the law of simple additivity over the range covered by these results, so that from known values for cokes produced from single coals, the structural parameters for the various blends can be calculated. The correspondence between experimentally determined and calculated values is shown in Table 1 for a series of cokes prepared from blends of four coals.

The value of the additivity lies in the ease with which predictions can be made about cokes produced from blends. For example for high coke strength it is essential to have low porosity of the order of 50% and in the above series (Fig. 2) it is clear from the values for the cokes from individual coals that no blends of these coals containing 50% of the base coal carbonized under the conditions used in preparing this coke series would achieve the desired level of porosity for a high strength coke.

5. INFLUENCE OF INCORPORATION OF LOW-VOLATILE ADDITIVES INTO A BLEND

In general and in accord with the observations of the influence of blend rank in the previous section, additions of low-volatile coals into a blend produce systematic decreases in the porosity and pore size. Again however the magnitude of the structural changes depends on the nature of both the added coal and the base coal blend but on average the addition of coals of about 12.5 - 20% (dmmf) volatile matter to a blend reduces the coke porosity by about 1.5% and the pore size by about 6% per 10% addition of the low volatile coal. For coals of less than 12.5% volatile matter the decrease in porosity and pore size are of the order of 3.7% and 5%, per 10% addition, respectively. There is however an additional effect observed with the lower volatile coals, of little or no dilatation, which produces an increase in the pore-wall thickness of the order of 12% per 10% addition. This can be compared with the previously observed increase in pore-wall thickness which resulted from adding delayed petroleum coke breeze to blends. For example, a petroleum coke of 14.7% volatile matter produced a 17% increase in the pore-wall thickness of coke produced from a coal of international class 634. In the latter case however, i.e. the addition of petroleum coke, no change in the coke pore size was found. Thus by selection of the appropriate low-volatile additives it is feasible to consider manipulating the porous structural features of the coke to achieve the required properties.

6. RELATION BETWEEN COAL DILATATION AND DEVELOPMENT OF COKE POROUS STRUCTURE

In the previous sections the emphasis was on the porous structure of the final coke product but it is also of interest to consider the process by which the porous structure develops and of the influence on this development of the coal properties. To this end, an examination of the effects caused by changes in the dilatation and volatile matter content of coal charges has been made using laboratory scale carbonizations.

A single-heated-wall oven was used to provide a carbonized block across which a temperature distribution of from 1000°C to 300°C is attained. By examining the products at different carbonization temperature levels it was shown that for dilating coals both the pores and pore walls attain transient maximum values within the highly porous zone formed in the plastic temperature range. However when the mean pore and pore-wall dimensions (as given by intercept counts) for the highly porous product and the coke, were compared with the total dilatation (as measured by the Ruhr dilatometer), the mean pore size in the highly porous zone showed a broad relation to the dilation but the structural data for the cokes showed no obvious relation with the dilatation behaviour, Fig. 3.

The method used does not lend itself to obtaining precise data due to the small number of pores along each isothermal traverse of the carbonized block and to overcome this limitation use was made of the tubular furnace which yields sufficient coke sample to enable statistically reliable values to be made (although it does not readily provide information on the course of the pore structure development process).

For a series of 15 cokes prepared from 3 coals blended to give specific volatile matter contents, there was no evidence for any systematic trends between the porous structure and the total dilatation, either for the individual blends of specific volatile matter or when considered as a whole, Fig. 4, and on the basis of this evidence the dilation as measured does not appear to exert a dominating influence on the pore structure of the final coke product.

7. SIGNIFICANCE OF THESE RESULTS IN RELATION TO INDUSTRIAL PRACTICE

The measurement of features of the coke porous structure can now be achieved with significant accuracy and precision to enable consideration to be given to the formulation of coal blends to give coke of specific structural characteristics.

By utilizing data such as that described in this paper the influence of different components in a coal blend can be ascertained and by appropriate selection of the blend components, the blend composition and the carbonization conditions, coke can be produced to meet particular requirements. The data required can generally be obtained from 250 or 350 kg oven carbonizations of the individual coals and the application of knowledge of the influence on the structure of changes in the various carbonization variables, although confirmatory tests on the blends using large scale ovens is considered a wise precaution at the present state of knowledge.

As an example of this method, a formulation was calculated whereby a caking coal of international class 634/5 could be used in a blend which would produce a high strength coke of specific structural characteristics. The coke from the caking coal had a porosity of nearly 61% with a pore size of 210 µm and a pore-wall size of 126 µm. For the specified high strength it was necessary to transform these values into about 47.6% porosity, 125 µm pore size and 140 µm pore-wall size. It was calculated that chese changes could be achieved by the use of preheating and the addition of 35% of a prime-coking coal (class 534/5), 20% of a low volatile coal (class 333/4) together with 20% of petroleum coke breeze. This blend was carbonized in the BCRA 6.5 m (17-t) test oven and the coke produced was within 1.2% of the specified porosity, 4 µm of the pore size and 10 µm of the pore-wall size. Consequently, the coke had the desired high strength properties required.

This example serves to demonstrate how the structural data can be used to formulate coal blends to produce coke of specified structural, and thereby strength, characteristics, but it is evident that the example gave one solution out of several possible alternative ways of achieving the desired structural changes. For a complete solution much more information is required about the influence of the various carbonization variables and our work is proceeding on these lines.

It should be emphasized that the foregoing refers to cokes with well bonded material in the pore-walls, i.e. cokes which are normally encountered in the metallurgical industry, and the various trends established are restricted in their range of application by this proviso. When poorly bonded material is present in the pore-walls some of the structural measurements are equivocal and further work is required to overcome this problem which is generally confined to cokes of low strength characteristics.

Some of our results have also highlighted the problems which arise when trying to use coal properties to predict coke properties. Thus the study of the influence of blend 'rank' as assessed by the reflectance of the vitrinite clearly shows how other factors have to be considered for a complete picture. The pore structure development would be expected to be a function of the dilatation and in the early stages of the pore development there is some evidence for this, but the structure of the final carbonized product does not appear to be strongly dependent on this particular coal property. Hence two of the coal properties most widely used in predicting coke quality, i.e. vitrinite reflectance, or volatile matter and dilatation, are inadequate as means of predicting coke porous structure which itself governs the intrinsic strength of coke.

ACKNOWLEDGEMENTS

The financial support of the European Coal and Steel Community is gratefully acknowledged.

REFERENCES

1. Patrick, J.W., Sims, M.J. and Stacey, A.E., J.Phys. D:Appl. Phys., 1980, 13, 937
2. Juranek,G., Bussman, B. and Seifert, G., Brennst.-Chem., 1963, 44, 97
3. Mackowsky, M-T and Wolff, E-M., 'Coal Science', Am. Chem. Soc. Advances in Chemistry Series 55, Washington DC, 1966, pp 527-548
4. Patrick, J.W., Sims, M.J. and Stacey, A.E., J. Microsc., 1977, 109, 137
5. Patrick, J.W., Paper to the ECSC 6th Round-table Meeting, Luxembourg, 1980.
6. Patrick, J.W. and Stacey A.E., Proc. 4th London Carbon & Graphite Conf. Soc. Chem. Ind., London, 1976, pp 634-643.
7. Hays. D., Patrick, J.W. and Walker, A., Fuel, 1976, 55, 297
8. British Carbonization Research Association, Carbonization Research Report 87, 1980
9. Inter alia, Foch, P. 'Coke in ironmaking' Iron & Steel Institute, London, Publication 127, 1970, pp 47-52.
10. Graham, J.P., Pater, W.J. and Lee, G.W., 'Coke in iron making', Iron and Steel Institute, London, Publication 127, pp 53-62

TABLE 1. COMPARISON OF MEASURED AND PREDICTED STRUCTURAL VALUES FOR A SERIES OF COKES

Porosity, %		Pore intercept size, µm		No. of pores/field	
Measured	Calculated	Measured	Calculated	Measured	Calculated
55.7	55.1	164	161	109	110
52.4	52.8	143	146	123	119
51.6	51.6	141	138	128	130
49.5	50.5	128	130	146	140
57.1	55.1	163	166	110	112
51.2	52.8	145	148	134	126
54.7	52.9	148	152	123	122
55.7	54.4	161	160	123	114

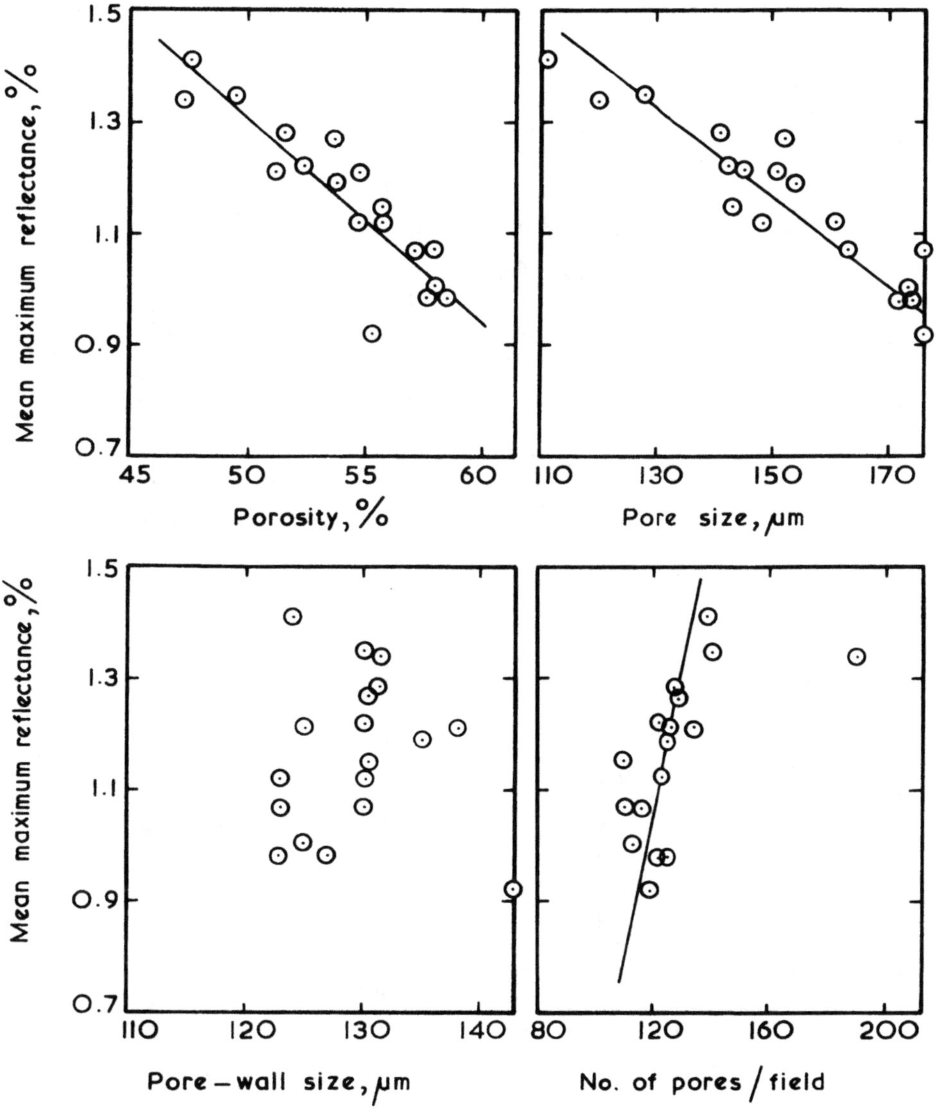

FIG. I. INFLUENCE OF COAL BLEND 'RANK' ON COKE POROUS STRUCTURE

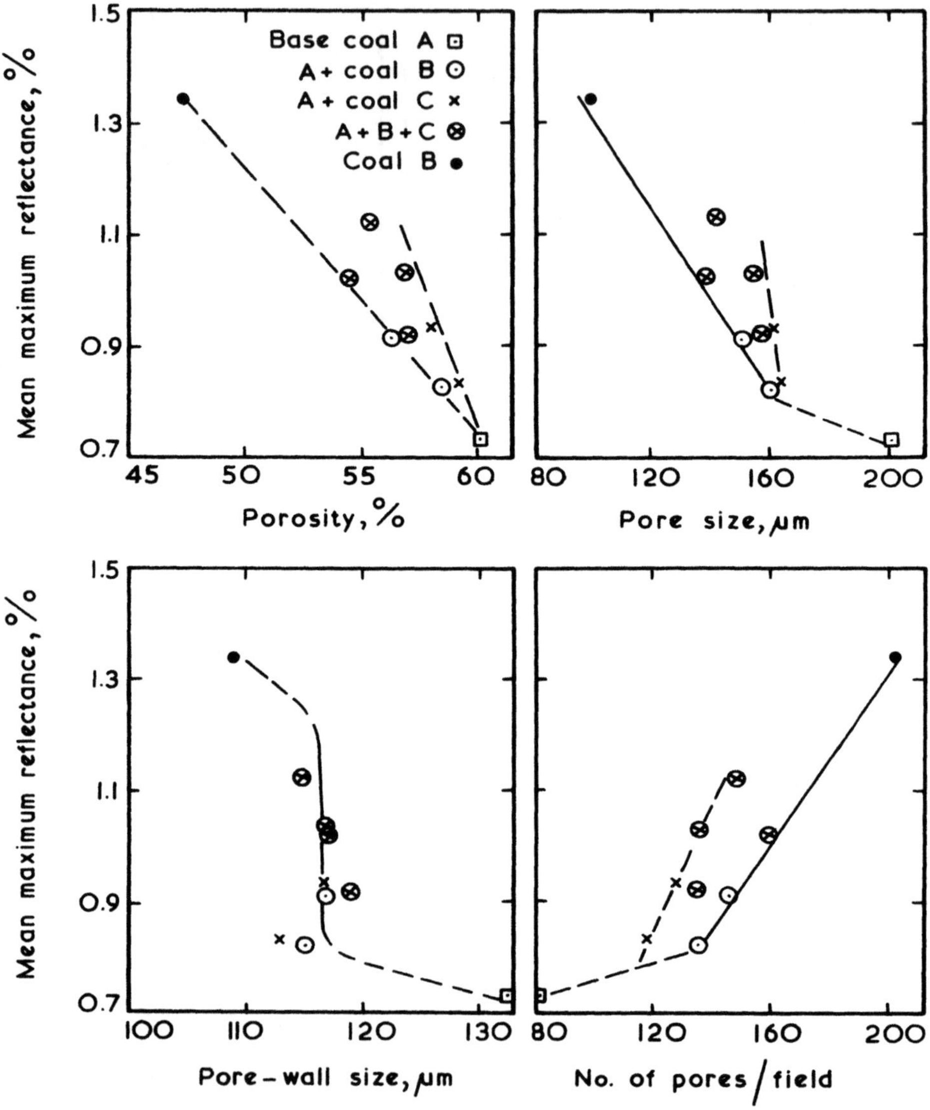

FIG. 2. INFLUENCE OF BLEND 'RANK' AND COAL TYPE ON COKE POROUS STRUCTURE

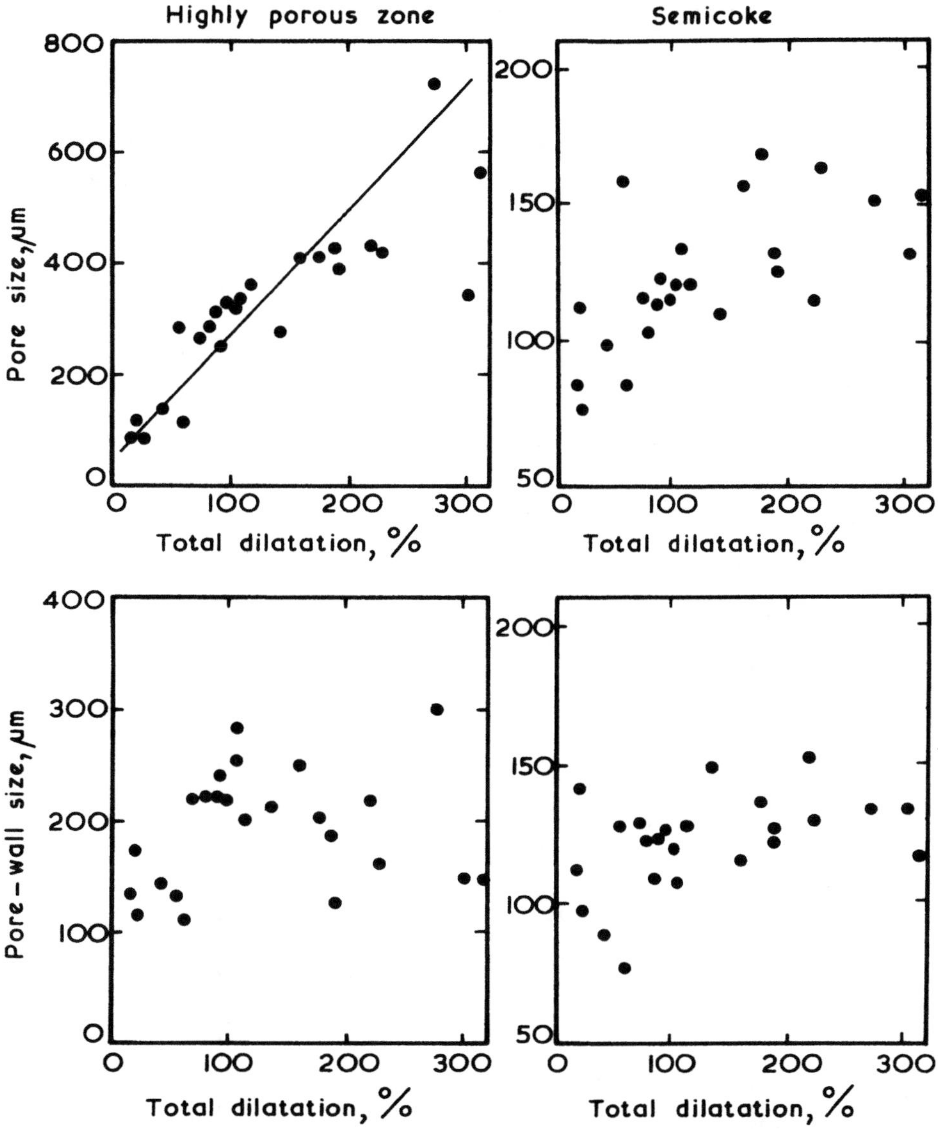

FIG. 3. VARIATION OF STRUCTURAL PARAMETERS FOR THE HIGHLY POROUS CARBONIZATION PRODUCT AND SEMICOKE WITH DILATATION

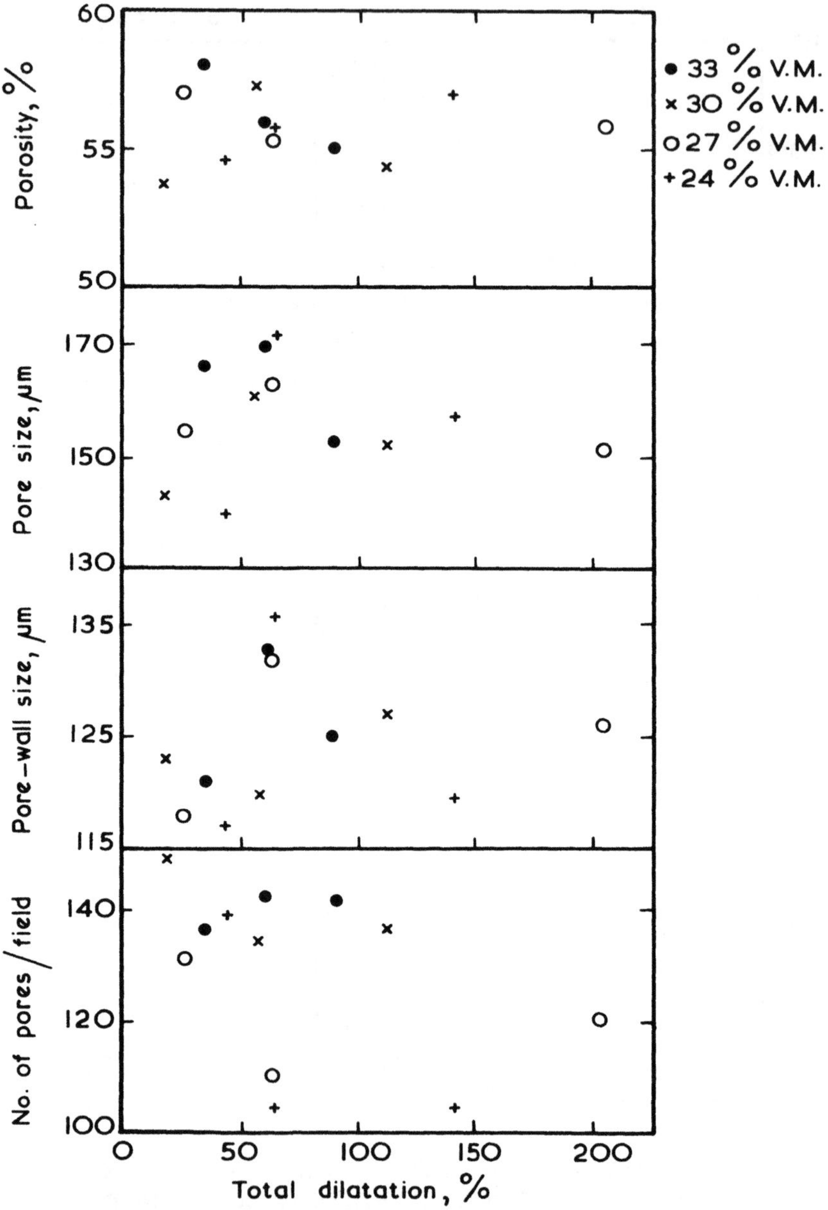

FIG. 4. VARIATION OF PORE STRUCTURAL PARAMETERS WITH BLEND DILATATION : LABORATORY SCALE CARBONIZATION

COAL CARBONIZATION: FORMATION, PROPERTIES AND RELEVANCE OF MICROSTRUCTURES IN RESULTANT COKES

H. MARSH

Northern Carbon Research Laboratories, School of Chemistry, University of Newcastle upon Tyne, Newcastle upon Tyne NE1 7RU, England

Summary

Metallurgical coke is a composite material which when used in the blast furnace has to meet several requirements which are not totally complementary. It is considered that metallurgical coke is a successful compromise in structure and properties. Initially, the microstructures or optical textures of cokes are described. They mainly consist of anisotropic mosaics. Anisotropic carbon in cokes is formed from the homogeneous nucleation of liquid crystals and mesophase from the melt. Factors, physical and chemical, which control the size and shape of optical texture are summarized leading to a discussion of optical textures of cokes from coals of increasing rank. The optical texture of mosaics is relevant to industrial performance. The interlocked, randomly orientated units of the mosaics are more resistant to fracture than isotropic carbon or the more orientated petroleum cokes. Anisotropic carbon is more resistant to gasification than isotropic carbon. The mosaics are less able to develop troublesome gasification fissures than the cokes of larger optical texture. The shrinkage fissures of the mosaics created by heat treatment permit some accommodation of internal stresses set up by thermal expansion. The mosaics are more resistant to alkali than the larger optical texture. Co-carbonizations are described which produce cokes with suitable optical textures. The concepts of 'hydrogen shuttling' is introduced to explain the successful use of pitch additives in coal blends.

1. INTRODUCTION

Metallurgical coke is used in the blast furnace as an energy source, a chemical reducing agent and to provide permeability and support for the furnace load. To fulfil these functions and to maintain blast furnace performance the coke must maintain its size within an optimum range. The cokes must therefore, keep its mechanical strength by withstanding as far as possible, in the blast furnace, degradation due to gasification in carbon dioxide, abrasion, compressive forces, thermal shock and reaction by alkali. In most countries metallurgical cokes are prepared from blends of coals sometimes with additions of pitch. The technology of coke making is established (1, 2) and specifications for metallurgical coke exist (3). Within recent years significant advances have been made in understanding the fundamental science of coke making. The purpose of this paper is to review some of these advances, in particular the contribution of the Northern Carbon Research Laboratories, in our knowledge of the mechanisms of formation of microstructures in coke and the relevance of these to coke properties.

2. MICROSTRUCTURES IN COKES

Metallurgical coke can be studied by reflected polarized light optical microscopy. This is done by preparing optically flat surfaces of a piece of coke by first cutting a piece of coke and polishing the cut surface using progressively finer grades of emergy paper and alumina powder (4). Examination of this polished surface by polarized light optical microscopy using a half-wave retarder plate placed between the specimen and the analyzer, shows immediately that the metallurgical coke is porous and that the solid pore-wall material reveals a composite structure.

The strength of a brittle porous material such as coke is strongly dependent on its porosity, the strength decreasing with increasing porosity. Aspects of this very important consideration are adequately covered by Patrick and Wilkinson (5).

For a coke of given porosity, its strength must be dependent upon the composition of the solid pore-wall material. Hence its detailed examination becomes very relevant. The optical microscope reveals that pore-wall material or coke substance is a multicomponent system. Generally, metallurgical cokes contain significant proportions of anisotropic, graphitizable carbon as distinct from isotropic, non-graphitizable carbon. This is revealed in optical microscopy by the colours generated by the half-wave retarder plate and by the interchange of colours which occurs on rotation of the specimen-stage (6). The anisotropic carbon is coloured blue, yellow or purple.

This anisotropic carbon exists as an agglomerate of units of size from < 0.5 μm diameter to ~ 50 μm. These units comprise what are termed the microstructure of cokes. The combined effect is to produce the optical texture of the coke. The anisotropic units can realistically also be called the macrocrystallites of the coke. It is within these macrocrystallites that the crystallites exists as described and measured by Blayden, Gibson and Riley (7) using X-rays.

It is the shape and size of the macrocrystallites, the method by which they are held together, how they interact with mineral matter and organic inerts within coal and their formation, which are discussed in this paper.

3. LIQUID CRYSTALS AND MESOPHASE

The formation of anisotropic carbon from the isotropic melt of coal, pitch or selected model organic compounds was first attributed to the development of a distinct phase of liquid crystals by Brooks and Taylor (8).

Some pure organic systems do not pass directly from a solid to a liquid on heating but do so via distinct phases, called liquid crystals, having properties intermediate between solids and liquids. Unlike a liquid, in a liquid crystal phase there is well-defined order on a length scale of ~ 10^2 μm, the fluid properties also being well-defined (9). There are several types of liquid crystal, nematics, smectics and cholesterics of which the nematics have immediate relevance to coal carbonization systems. The conventional nematics are composed of rod-like molecules with strong polarity but without facility for intra-molecular rotation. The long axes of these rods are aligned about a common direction in space but the centres of mass of the rods are randomly distributed such that no long-range order exists and an X-ray diffraction would show no Bragg peaks.

The isotropic melt of the carbonization sytem forms lamellar nematic liquid crystals as distinct from the rod-type. A carbonizing or pyrolyzing system about to produce liquid crystals (630-700 K) contains aromatic molecules with a large spread of molecular weights. These molecules interact chemically and polymerise via cross-linkages using dehydrogenative polymerization reactions (10). This polymerization creates planar molecules consisting of sheets of hexagonally-linked carbon atoms containing holes, heteroatoms and free electrons. When the molecular weight is > ~ 1000 amu then the process of homogeneous nucleation creates the association of these molecules into the nematic liquid crystal. Inter-molecular van der Waals' forces maintain the stability, initially, of these liquid crystals. The process of homogeneous nucleation is reversible during the very early stages of formation (11) but as the carbonization progresses so further chemical cross-linkage reduces this fluidity. This polymeric phase is termed mesophase and is the intermediate phase between the isotropic fluid coal and the solid anisotropic semi-coke ultimately formed from the mesophase. Thus, the mechanism of formation of the ordered anisotropic carbon is not one of recrystallization or of precipitation but is of homogeneous nucleation of fluid liquid crystals. Those factors which influence this nucleation process and resultant fluidity of the nematic liquid crystals will also influence the appearance or optical texture of the resultant anisotropic coke.

4. OPTICAL TEXTURE OF COKES

Not only metallurgical cokes but cokes from coal-tar and petroleum pitch and model organic compounds exhibit optical textures. The microstructures in these cokes can vary in size from < 0.5 μm diameter to > 200 μm in length. In order to understand how these variations are brought about it is important to appreciate that the carbonization process is usually non-isothermal. Unlike the conventional liquid crystals (9) described above these lamellar nematic liquid crystals are created in a highly reactive pyrolyzing system which is not a system at equilibrium. Mesophase is formed continuously over a range of increasing temperature. With increasing temperature two competitive processes occur. One is of continued cross-linkage within the mesophase the effect being to decrease the fluidity of the mesophase. A second is the normal effect of increased fluidity associated with the increasing temperature of the system.

Fluidity is of major importance in determining the shape and size of resultant optical texture of cokes. This is because the liquid crystals, when forming from the isotropic fluid phase of carbonization, grow in size and ultimately touch each other. If the fluidity is sufficiently high it is observed that, on touching, the liquid crystals (often spherical in shape) coalesce immediately into a single larger unit and so the process continues. If the fluidity is not sufficiently high then the liquid crystals (or anisotropic growth units) cannot coalesce but retain their identity. The optical texture of a coke is thus determined by the extent of coalescence of mesophase which occurs during the carbonization process. In addition, mechanical deformation of mesophase caused by bubble percolation or convection currents create flow-type anisotropic carbon. Thus the optical texture of carbons is made up of microstructures (macrocrystallites) and these need to be defined as suggested in Reference 19.

5. SIZE AND SHAPE OF OPTICAL TEXTURE

Factors which influence the size and shape of optical texture are reviewed by Marsh (12), Marsh and Cornford (13), White (14), White and Zimmer (15) and Marsh and Walker (16). In general terms it is found that aromatic systems, such as from anthracene, form cokes with large sized optical textures of flow domain anisotropy and domains, > 60 μm diameter. The presence of heteroatoms and functional groups in the reacting molecules generally brings about a decrease in size of optical texture. This is because cross-linkage is facilitated between constituent molecules at too early a stage in the growth of mesophase. This in turn creates a system which has too low a fluidity such that the growth units of mesophase cannot coalesce but remain small and so create the optical textures of mosaics, 0.5 to 10 μm diameter. The ultimate in this process is the formation of isotropic carbon where cross-linkage occurs prior to formation of liquid crystals.

6. OPTICAL TEXTURES OF COKES FROM COALS OF DIFFERENT RANK

There exist very significant differences in the size and shape of optical textures of cokes from coals of different rank. Patrick et al. (17), Marsh et al. (18) and Grint et al. (19) demonstrated, using several experimental approaches, that the size of optical texture decreases with decreasing coal rank. The high rank low volatile anthracites and semi-anthracites carbonize without significant softening such that the resultant 'coke' is highly anisotropic because of the preservation of the inherent anisotropy of the parent coal. The prime coking coals produce cokes with the largest size of optical texture ranging from very fine-grained mosaics, < 0.5 μm diameter to medium-flow anisotropy, < 30 μm in length, < 5 μm in width. The cokes from low rank high volatile weakly caking coals contain perhaps about 25% of anisotropic carbon with an optical texture of fine-grained mosaics, 0.5 to 1.5 μm diameter. The low rank non-caking bituminous and sub-bituminous coals (volatile matter > 40%) produce cokes which contain only isotropic carbon.

Explanations for these variations in size and shape of optical texture can be given initially in terms of the maximum fluidity of the carbonization system and the range of temperature of fluidity (20). Generally the maximum fluidity as well as the temperatures of maximum fluidity increase with rank of coal. This temperature of maximum fluidity can increase from ~ 693 K (~ 420°C) to 753 K (480°C) volatile matter 40% to 20%. Likewise, the range of temperature of fluidity can increase from ~ 60°C to ~ 160°C. The physical properties are in turn controlled by the chemical properties of the carbonizing system. The heteroatom content and functionality of coals decrease with increase in rank. Consequently as the coal material becomes more aromatic and more polymerized as a result of coalification so the overall chemical reactivity of the system decreases. This permits an increase in temperatures of maximum fluidity, of maximum fluidity itself and in the range of temperature of fluidity. These chemical changes associated with the optimization of physical conditions for development of liquid crystals (aromatic molecules and high fluidity at relatively high temperatures) bring about the optimization of extent and size of anisotropic carbon from mesophase in cokes from coals. With the higher rank anthracites, the extent of polymerization of the coal prevents thermolysis, the coal never becomes fluid and no anisotropic carbon is created via the mesophase route.

Thus, the coking process of the coking coals represents an optimization of several events. The coalification process is not so advanced as to prevent thermolysis and fluidity. The extent of aromatization is sufficiently advanced to permit the formation of aromatic lamellar nematic liquid crystals. The thermal stability of the molecular constituents of the carbonizing system is sufficiently low so as to prevent extensive loss of volatile matter. This subsequently prevents the formation of a 'frothy' or highly porous coke. Instead a coke of optimum porosity (5) is created, an additional factor being the loss of volatile matter during the period of decreasing fluidity of the system with increasing carbonization temperature. The residual heteroatom content and functionality of the coking coals permits the formation of the mosaics of optical texture (0.5 to 10 μm diameter) and some medium-flow anisotropy (< 30 μm in length). Should the heteroatom content and functionality have decreased further, and if the same extent of thermolysis was possible, then cokes resembling the petroleum derived cokes as used in graphite manufacture would have resulted. Clearly these would be quite unsuitable for the blast furnace.

7. PROPERTIES AND RELEVANCE OF OPTICAL TEXTURE

It is opportune to ask if the optical texture of metallurgical cokes, composed as it is of anisotropic mosaics, some flow-type anisotropy, mineral matter and inert organic matter represents an ideal structure for such a composite material. Or, posing the question another way, can the properties of a good metallurgical coke be explained in terms of its optical texture in association with its porosity (5). It is not unreasonable to say 'Yes' to both of these questions.

7.1 COKE STRENGTH

In a porous solid such as coke, considerations of strength must invoke discussions of crack generation, propagation and stoppage. In isotropic materials brittle-fracture can occur to create conchoidal fracture surfaces, e.g. glassy carbon. In anisotropic carbon with large optical texture it is relatively easy for crack propagation to occur parallel to the basal planes of constituent lamellae, e.g. needle-cokes from petroleum pitch. Metallurgical coke is a heterogeneous composite of anisotropic mosaics and this may be an advantage. Three aspects can be considered.

A study (21) of the carbonizing system has been made at an intermediate stage prior to complete formation of semi-coke. The systems were thus a mixture of uncarbonized coal and mesophase. Surfaces of this mixture were washed with chloroform to reveal the porosity. Optical microscopy of cross-sections of this 'washed' systems revealed layers of anisotropic material forming a 'lining' on the surfaces of these pores. This is equivalent to a large increase in surface tension of the fluid phase surrounding the pore. This in turn would restrict pore growth and prevent the formation of 'frothy' coke. Such a finding is commensurate with the knowledge that liquid crystals tend to flow such that the lamellar molecules align themselves parallel to surfaces (22).

A study (21) has been made of the effect of inert material of small size upon the growth and coalescence of mesophase. Ashland A240 pitch and a coal-extract were co-carbonized with Carbolac 2, a carbon black of particle diameter of 100 nm. The more fluid mesophase from pitch was able to disperse the carbon black, to prevent its adsorption on to surfaces of mesophase and hence to maintain the large size of optical texture. The less fluid mesophase from coal-extract could not do this and a smaller size of optical texture resulted. Thus in coal carbonizations, the finely dispersed mineral matter and inert (non-reactive) organic components will tend to prevent the formation of larger sized anisotropic optical texture in the resultant coke.

A study (23) has been made of the structure of the carbon located at the interface between the individual units of fine-grained mosaics of a metallurgical coke. A polished surface of such a coke was etched with atomic oxygen and the resultant topography examined by scanning electron microscopy. The carbon which constitutes the interfacial position was etched preferentially to reveal pentagonal cross-sections of sizes equal to those of the fine-grained mosaics. This preferential etching could occur because the carbon at the interface is more disordered (even isotropic) than the carbon of the anisotropic fine-grained mosaics. It has also been observed that the growth units of mesophase from coal are not truly spherical but are polyhedral (18). The etching studies suggest that although the fluidity of these growth units of mesophase is too low to facilitate total coalescence, they are sufficiently plastic to have their shape moulded into pentagonal cross-sections by the internal compressive forces existing during the final stages of the fluid phase of carbonization. Thus, these compressive forces disturb stacking sequences of basal planes at surfaces of the growth units and so create boundary layers of disordered carbon which 'interlock' these growth units so enhancing their cohesion. Accordingly there does not exist a smooth interface between the units of the mosaics, and hence crack propagation is not facilitated.

A comparative study of the microstrength of cokes and their optical textures has been made by Ragan and Marsh (24). These authors suggest that increases in strength of the pore-wall material of several cokes are associated with the development of optical textures of mosaics as these tend to resist propagation of cracks which may originate, for example, in the domain anisotropy of coke from one coal component in the carbonizing blend.

Thus it would appear that the operational strength of pore-wall material of metallurgical coke results from the composite structure of anisotropic mosaics. In this type of material resistance to crack propagation may be maximized.

7.2 COKE REACTIVITY

It is important in discussions of coke reactivity to consider both the rates of gasification (under standard conditions) and the surface topography induced by gasification, i.e. the topochemistry of the system.

Blast furnace operation requires a minimum gasification of coke prior to entering the race-way. Gasification in the upper regions of the blast furnace creates 'solution-loss' which is uneconomic (25). Further, the development of macrocracks or fissures by highly selective gasification mechanisms results in a very marked decrease in strength of cokes (mainly petroleum type cokes).

It is established that anisotropic carbon is less reactive to carbon dioxide than is isotropic carbon (26). There may be two reasons for this: a lower surface area of the carbon available for reaction; the intrinsic reactivity may be lower for 'molecules' which constituent anisotropic carbon. Hence, the formation of anisotropic carbon in metallurgical coke minimises reactivity from a structural viewpoint. See Figures 1 and 2.

In a study of gasification of polished surfaces of metallurgical cokes, Forrest and Marsh (27) noted that the size, shape and orientation of the gasification fissures were typically those of the anisotropic mosaics. The fissures are created in such a way as to minimise the facility for crack propagation. The orientation of the fissures is parallel to the basal planes of the units of the mosaics and hence it is random over the coke surface because the growth units of the mesophase which lead to mosaics are themselves randomly orientated. The flow type anisotropy gasifies to create fissures which have the dimensions of the length of these units. As such they are less resistant to crack propagation.

Thus, it would appear that metallurgical coke is a successful compromise between the low reactivity of anisotropic carbon and the fissuring tendency of the larger sized microcomponents of optical texture of anisotropic carbon.

7.3 RESISTANCE TO THERMAL SHOCK

Forrest and Marsh (27) heated three cokes to 2073 K under nitrogen and monitored the development of shrinkage fissures with direct reference to optical texture. This treatment opened up small fissures of typical size 1 to 5 μm, mainly in the mosaics. These fissures were orientated parallel to the larger planes of the anisotropic carbon. The fissures cease abruptly where there is a rapid change in orientation of, or discontinuity in, these planes, as seen as a change in colour under the optical microscope. Hence, fissure propagation seems to be restricted by the frequent rapid changes which occur in structural orientation. A few larger fissures, of typical size 30 x 5 μm occur in the flow-type anisotropy. Thus, it is possible to suggest that the fissures created by shrinkage of coke, accompanying densification resulting from increased heat treatment temperature, permit a significant accommodation of internal stresses set up by the thermal expansion of the coke. The random orientation of the shrinkage fissures favours the maintenance of the coke strength.

7.4 RESISTANCE TO REACTION WITH ALKALI

In a comparative study of cokes taken via the tuyeres with cokes of original feedstock, Monson and Wilkinson (28) indicate that the changes which occur in the properties of cokes in the blast furance are the result of both thermal and chemical factors. Previously it had been thought that the coke degradation resulted almost entirely from mechanical action. The presence of alkali in the blast furnace (29) can have a two-fold deleterious effect. It can catalyse the solution-loss reaction and also, in an inert atmosphere, can cause disintegration of the coke substance. Marsh and Wilkinson (30) studied the interaction of potassium hydroxide with several cokes of increasing content of anisotropic carbon. Polished surfaces of cokes were prepared. The alkali was placed on to these surfaces from aqueous solution and the system heated to 1273 K (1000°C) under nitrogen. The topography created by the reaction of the alkali was monitored by scanning electron microscopy and compared directly with the optical texture as recorded by photomicrography. It was observed that the extent of reaction of potassium hydroxide with metallurgical coke increased with increasing size of the anisotropic unit of optical texture. For example, the flow-type anisotropy reacted preferentially to create extensive fissuring. Further, the reaction with potassium hydroxide decreased the microstrength of the cokes, the effect increasing with increasing content of anisotropic carbon. See Figure 3.

Thus, it would appear that metallurgical coke with its optical texture predominantly of mosaics is a natural compromise between isotropic cokes (less reactive to reaction with alkali) and the anisotropic cokes with flow-type anisotropy (more reactive to reaction with alkali).

8. CO-CARBONIZATIONS TO PRODUCE COKES WITH SUITABLE OPTICAL TEXTURES

Currently, it is common practice to produce metallurgical cokes from blends of coals of different rank (2). Considerable care is exercised in blend formulation because of the limited nature of the process (2, 31). Little or no interaction occurs between prime coking and higher rank coals (32). A strong interaction occurs between prime coking coals and the lower rank medium and high volatile caking coals. Little interaction occurs on co-carbonization of prime coking coals with the weakly caking high volatile coals. Where interactions occur it is observed that the optical texture of resultant coke is clearly not that of either of the coal-cokes but is different and is developed from the fluid blend of the two parent coals. The prime coking coal up-grades the size of the mosaics beyond that observed in cokes from the lower rank coal of the acceptable blend.

Within recent years attention has been given to the use of petroleum and coal-based pitch-like products to replace prime-coking coals in blast formulations as described for example by Nakamura et al. (31) in the use of briquettes. The Northern Carbon Research Laboratories have made a systematic study of the use of such materials in blend formulations.

Initially using model compounds, Mochida et al. (32, 33) studied the co-carbonization of a range of coals with acenaphthylene and decacyclene. the latter compound being extremely able to modify the carbonization or coals of most ranks. Grint et al. (19) extended the study by using Ashland A200 petroleum pitch as additive and noted the ability of this pitch to interact with most coals. Similar effects, but not so extensive, were observed when solvent refined coals and coal-extracts were used as additives (34, 35, 36). It was then observed that hydrogenated derivatives of petroleum pitches and coal-extracts were superior modifiers than the non-hydrogenated materials (37, 38). Increasing the rate of carbonization resulted in an increase in size of optical textures of cokes from the co-carbonization of coals with Ashland A240 petroleum pitch (39), a result in keeping with the known effects of heating rate on fluidity (40).

In an examination of the co-carbonization of one coal with several industrially supplied petroleum pitches, Grint and Marsh (41) noted that the pitches differed considerably in their ability to modify the carbonization behaviour of the coal. Some pitches (passive) had little effect on the coal, the product being a composite of coal-coke and pitch-coke. With the active pitches, the latter interacted totally with the coal to produce a coke of different optical texture, no pitch-coke being detectable in the resultant coke composite. These authors then demonstrated with laboratory prepared cokes and with cokes prepared in a 7 kg oven that a significant increase in coke strength could be associated with the use of pitch additives in coal blends (42). As mentioned above, the analysis of the optical textures of cokes from coal blends and coal/pitch blends demonstrated a correlation between extent of formation of mosaic type anisotropy and the microstrength of the coke (24, 42).

9. THE CHEMISTRY OF CO-CARBONIZATIONS TO PRODUCE ACCEPTABLE COKES

As the chemical compositions of coals and pitches are extremely complex so no exact explanation exists of differences in size and shape of optical texture of resultant cokes in terms of the pyrolysis chemistry of the carbonization process (43). However, it is known that pitch materials differ widely in their ability to modify coal carbonization systems (41) so some explanation must be sought. It has been suggested that naphthenic groups and aromatic ring systems promote the growth of larger sized optical textures whereas the presence of heteroatoms and functional groups promote the growth of smaller sized optical textures (32). Recently, the concept of 'hydrogen shuttling' has been introduced in discussions of the development of mesophase during co-carbonization of coals with pitches (44). It is suggested that the hydrogen may stabilize certain of the free-radicals produced during pyrolysis (45). Larsen et al. (46, 47) consider coals to be good hydrogen donors and acceptors. The beneficiation of hydrogenation of coal additives is reported (38) as well as of the feedstocks used to make needle-cokes in delayed cokers (48).

Thus, the process of modification of carbonization processes by pitch additions is probably associated with hydrogen transfer reactions (hydrogen shuttling) involving movement of hydrogen from the pitch to radicals of the carbonization process. The stabilization of otherwise reactive radicals by hydrogen acceptance prevents formation, at too early a stage in the process, of coal-coke and allows the pitch to interact chemically with the coal ultimately to create a new beneficial optical texture via formation of nematic liquid crystals. Marsh, Yokono and Lopez demonstrated a correlation between modifying ability of pitches and hydrogen transfer capability using 9,10-dihydroanthracene (Unpublished results).

ACKNOWLEDGEMENTS

The author is appreciative of support from the ECSC 7220-EB-815 which made possible most of the studies reported in this paper and carried out in the Northern Carbon Research Laboratories. The editorial assistance of Miss Bridget A. Clow is acknowledged.

REFERENCES

1. Gibson J. and Gregory D.H., Coke Oven Managers' Yearbook, Mexborough U.K., 1978 p. 159.
2. Gibson J., Coke Oven Managers' Yearbook, Mexborough U.K., 1980, p. 246.
3. Grainger L., Coke Oven Managers' Yearbook, Mexborough U.K., 1975, p. 282.
4. Marsh H. and Smith J., Analytical Methods for Coal and Coal Products. Ed. Clarence Karr, Jr., Vol. II, Academic Press, N.Y., 1978, pp. 371-414.
5. Patrick J.W. and Wilkinson H.C., Coke Oven Managers' Yearbook, Mexborough U.K., 1977, p. 245.
6. Forrest R.A. and Marsh H., Carbon, 1971, 15, 348.
7. Blayden H.E., Gibson J. and Riley H.L., "Coals, Cokes and Chars", B.C.U.R.A., 1943, p. 176.
8. Brooks J.D. and Taylor G.H., Chemistry and Physics of Carbon, Ed. Walker P.L. Jr., Marcel Dekker, N.Y., 1968, 4, 243.
9. Priestley E.G., (Ed.) "Introduction to Liquid Crystals", Plenum, N.Y., (1975).
10. Evans S. and Marsh H., Carbon, 1971, 9, 733.
11. Lewis R.T., Ext. Abs., 12th Biennial Conf. on Carbon, Pittsburgh (1975), Am. Carbon Soc., p. 215.
12. Marsh H., Proc. 4th London Inter. Carbon and Graphite Conference, SCI, 1976, pp. 2-38.
13. Marsh H. and Cornford C., "Petroleum Derived Carbons". ACS Symposium Series, Washington D.C., 1976, 21, pp. 266-281.
14. White J.L., ibid, pp. 282-314.
15. White J.L. and Zimmer J.E., "Surface and Defect Properties of Solids", Chemical Society, London, 1976, 5, pp. 16-35.
16. Marsh H. and Walker P.L. Jr., Chemistry and Physics of Carbon, Eds. P.L. Walker Jr. and P.A. Thrower, Marcel Dekker N.Y., 1979, 15, pp. 229-286.

17. Patrick J.W., Reynolds M.J. and Shaw F.H., Fuel, 1973, 52, 198.
18. Marsh H., Dachille F., Iley M., Walker P.L. and Whang P.W., Fuel, 1973, 52, 253.
19. Grint A., Swietlik U. and Marsh H., Fuel, 1979, 58, 642.
20. Lowry H.H. (Ed.) Chemistry of Coal Utilization, Supp. Volume, John Wiley, N.Y. 1963, p. 745.
21. French M.A., Ph.D. Thesis, Northern Carbon Research Laboratories, School of Chemistry, University of Newcastle upon Tyne, U.K., (1979).
22. Marsh H., Fuel, 1973, 52, 205.
23. Marsh H., Forrest M. and Pacheco L.A., Fuel, 1981, 60 (No. 5).
24. Ragan S. and Marsh H., Fuel, 1981, 60 (No. 6).
25. Strassburger J.H., (Ed.) "Blast Furnace Theory and Practice", Gordon and Breach Science Publishers, 1969, Vol. I, p. 68.
26. Tsuyuguchi M., Yamada T. and Sunami Y., Proc. 5th London Inter. Carbon and Graphite Conference, SCI, London, 1978, Vol. III, p. 8.
27. Forrest M. and Marsh H., Fuel, 1981, 60, (No. 5).
28. Monson J.R. and Wilkinson H.C. "Coal, Coke and the Blast Furnace", The Metals Society, London, 1978, p. 83.
29. Nakamura N., Togino Y. and Tateoko M., ibid, p. 1.
30. Marsh H. and Wilkinson A.P. Unpublished results.
31. Nakamura N., Togino Y. and Adachi T. ibid, Ref. 28, p. 93.
32. Mochida I. and Marsh H., Fuel, 1979, 58, 626; 803; 809.
33. Mochida I., Marsh H. and Grint A., Fuel, 1979, 58, 633.
34. Mochida I. and Marsh H., Fuel, 1979, 58, 790.
35. Marsh H., Mochida I., Macefield I. and Scott E., Fuel, 1980, 59, 514.
36. Marsh H., Kimber G.M., Rantell T. and Scott E., Fuel, 1980, 59, 520.
37. Mochida I. and Marsh H., Fuel, 1979, 58, 797.
38. Marsh H., Mochida I., Scott E. and Sherlock J., Fuel, 1980, 59, 517.
39. Marsh H., Gerus-Piasecka I. and Grint A., Fuel, 1980, 59, 343.
40. Ref. 20, p. 171.
41. Grint A. and Marsh H., Fuel, 1981, 60, (No. 6) Part 20.
42. Grint A. and Marsh H., Fuel, 1981, 60, (No. 6) Part 21.
43. Yokono T., Miyazawa K., Sanada Y. and Marsh H., Fuel, 1979, 58, 691.
44. Marsh H. and Neavel R.C., Fuel, 1980, 59, 511.
45. Petrakis L. and Grandy P.W., Fuel, 1981, 60, 115
46. Larsen J.W. and Sams T.L., Fuel, 1981, 60, 272.
47. Larsen J.W., Sams T.L. and Rodgers B.R., Fuel, 1981, 60, 335.
48. Sooter M.C., Becraft L.E. and Kegler W.H. U.S. Patent, 4,213,846, 22nd July 198

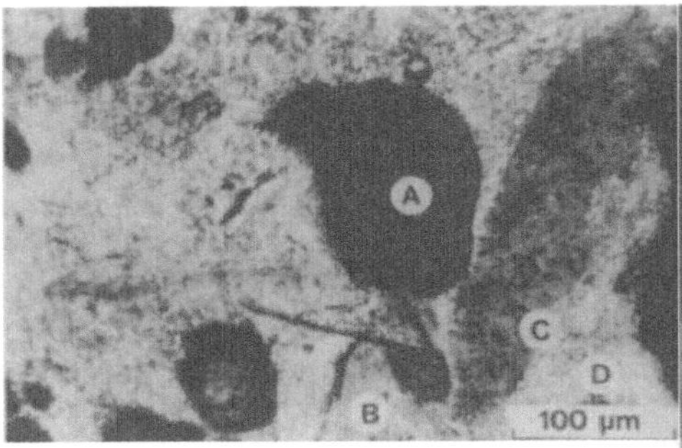

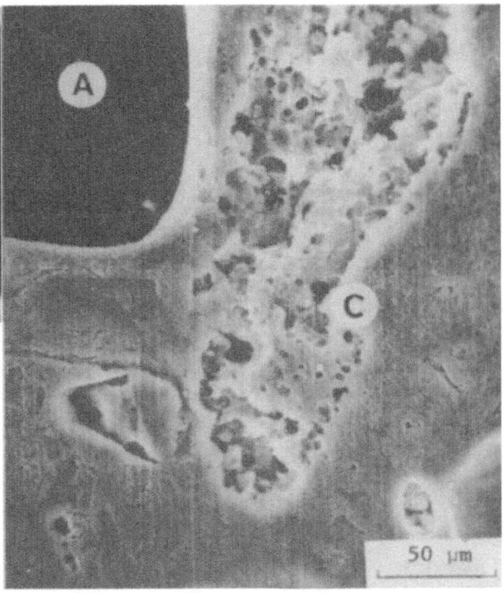

Figures 1,2. Optical and scanning micrographs of surface of coke from coking coal before and after gasification, CO_2, 1173 K, 1 h. Note preferential gasification of isotropic carbon at Position C.

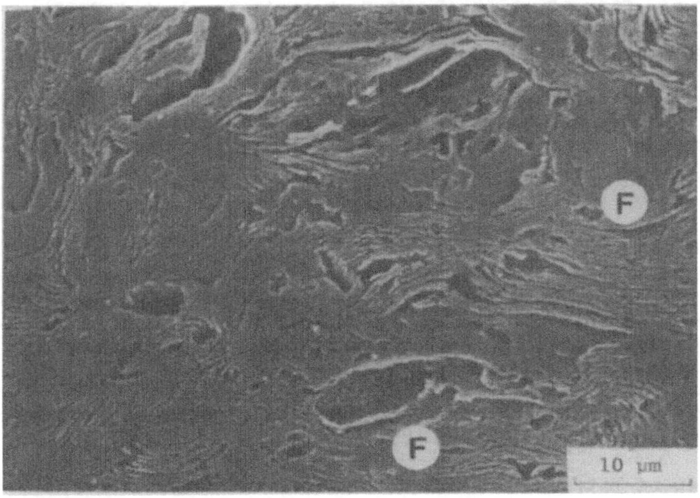

Figure 3. Scanning electron micrograph of surface of metallurgical coke after reaction with KOH, 873 K, 2 h. Note development of fissures in flow-type anisotropy, Positions F.

INFLUENCING OF THE PLASTIC PROPERTIES OF STEAM COAL BY MEANS OF BITUMINOUS ADDITIVES

O. Abel, E. Hopmann
Technological University of Clausthal

Summary

It is a well-known fact that when coal has a low gas content and inadequate coking capacity the coke quality can be considerably improved by adding binding agents such as pitch, bitumen or asphalt. It is generally assumed that the additive acts by increasing the plasticity of the "molten coal" during pyrolysis.

The present report provides results of laboratory studies of blends of steam coal and various types of carbo-pitch and petro-pitch as well as bitumen extract from the same coal. It examines the changes in:

- Gieseler plasticity
- dilatometer readings
- Roga Index.

In order to establish the effect of bitumen extracted from the same coal on the coking properties, the influence of both bitumen removal and bitumen addition is examined. The findings show that the pressure extraction treatment for obtaining the coal extract causes much greater changes in coking behaviour than actual bitumen removal.

Comparative tests with the oil bitumen and solid bitumen isolated from steam coal bitumen show that

- both types of bitumen increase dilatation, but oil bitumen is less effective than solid bitumen
- oil bitumen lowers the Roga Index whereas solid bitumen increases it.

1. Introduction

In the last few years there have been increased efforts to broaden the range of coking coals.

Major progress has been made towards this target with the preheating of high-volatile coal.

In the case of low-volatile coal with a low degree of fluidity and high inertinite content, the impact and abrasive strength of coke can be considerably improved by the addition of binding agents.

The binding agents which may be used include both coal-based products such as extracts (SRC) or tar and coal tar pitch, and oil-based products such as bitumen, oil pitch or asphalt.

It is generally believed that the binding agent affects the coal chiefly by increasing its fluidity and interacting with vitrinite so that the reactive proportion of the coal is increased and the inert macerals are more firmly bonded into the coke's structure.

Fig. 1 shows how the impact and abrasive strength of a steam coal are increased by the addition of tar pitch (1).

Studies of the correlations between the Gieseler plasticity of coking coal and coke strength carried out by Japanese experts, show that the Drum index DI_{15}^{30} varies almost linearly with the maximum Gieseler plasticity F_{max} within a range of 60-200 ddpm. In the range from 200 to about 1000 ddpm, DI_{15}^{30} remains almost constant and at higher values it falls again as a result of excessive plasticity (Fig. 2). The diagram also shows that coke strength is influenced not only by fluidity but also, to a critical extent, by the coal's carbonization degree. This is the basis for the MOF diagram (Fig. 3) in which coal types are rated and relected according to suitability for coke manufacture.

2. Aims of our experiments

In view of the importance of the controlled increase in fluidity for the coking of low-volatile coal and the cost of the binding agents required, the following questions in particular are of interest:

a) What bituminous additive provides the greatest increase in the maximum Gieseler fluidity F_{max}?

b) What are the effects of fluidity-promoting agents on dilatation and the G-value?

c) How do fluidity-promoting agents alter the Roga Index, i.e. the capacity of coal to bond inert substances?

d) What is the fluidity-promoting effect of the coal's own bitumen and what is the initial proportion of coal bitumen for changing dilatation and the Roga Index?

3. **Experimental methods**

The types of coal and bitumen/coal blends are characterized by the following properties or methods of analysis:

a) Coking capacity in accordance with DIN 51 739
Coking plant coal should normally have a dilatation of 30 to about 110.

b) Roga Index in accordance with ISO 335-1974
The Roga Index ranges from 0 to 100; coking plant coals should have a Roga Index of at least 45 and in general it lies between 55 and 70.

c) Maximum fluidity F_{max} in accordance with Geiseler, ASTM 2639-74.
In order to determine the Gieseler plasticity, the sample is compacted around the agitator in a measuring cell and heated up from 300 to 500°C at a rate of 3 K/min, whilst a torque of 0.01 Nm is applied to the agitator. The measurement result is the resultant angular velocity given in graduations per unit of time (1 rpm = 100 grads/min or 100 ddpm).

The initial softening can be traced much more clearly on the Gieseler Plastometer than on the dilatometer, but the softening temperatures differ considerably.

The most important parameter of fluidity progression is the maximum fluidity F_{max} and the temperature at that point, but the softening and resolidification temperatures and the extent of the plastic (Gieseler) region may also be useful parameters.

The binding agents and bitumen extracts from coal are identified by elementary analysis, mean molar mass and structural group analysis (calculated from IR and NMR analyses), as well as DTG analysis.

4. Experimental materials

4.1 Coal

The studies of the alteration of Gieseler fluidity by the use of bituminous additives were carried out mainly on steam coal with 16.3 % liqu. constituents (K16) and the studies on the effect of the coal's own bitumen on steam coal with 17.5% liqu. constituents (K17) and on bituminous coal with 22.3% liqu. constituents (K22). The analytical data of the tested coal types are given in Table 1.

4.2 Binding agents

The binding agents used for the experiments were two coal products (coal tar pitch, solvent refined coal (SRC)) and eight oil products of various qualities. Data available so far on the binding agents are given in Table 2.

4.3 Coal bitumen

Coal bitumen was obtained from various coking coals (K17, K22, K37) at decomposition temperatures of 260, 280 and 305°, followed by exhaustive extraction by the use of benzene in a Soxhlet apparatus. Part of the bitumen was cut into petroleum ether, oil bitumen and solid bitumen (insoluble in petroleum ether) by means of precipitation (3). The analytical data are given in Table 2.

5. Results

5.1 Effect of binding agents

As can be seen from Fig. 4, whatever the origin of the bituminous binding agents used, their blending causes a roughly linear increase on a logarithmic graph in the F_{max} value of the binding/agent/coal blend. Nevertheless, the coal-based binding agents (carbo-pitch, SRC) are evidently much more suitable than the oil-based binding agents.

In order to achieve maximum fluidity rates of 60 to 1000 ddpm (considered optimum for the Drum Index according to the studies of the Japanese authors) (2) with K16 steam coal, quite substantial amounts of additives are necessary, irrespective of the type used. Even the very "active" carbo-pitch has to be added at a rate of about 9% to obtain the fluidity range in question. The proportion required in the case of SRC, which is also quite suitable, goes up to 12%. If quantities of binding agent are to be restricted to a tolerable limit, the only additives which may be blended with K16 steam coal to obtain the required maximum fluidity are carbo-pitch and SRC.

The situation is basically similar for the more highly volatile K17 steam coal (Fig. 7). Here, too, carbo-pitch is much more effective than the best of the oil bitumens: No 1. The critical lower limit of maximum fluidity of 60 ddpm, however, can be reached with much smaller quantities of binding agent in the blend; this also applies to blends with bituminous coal K22.

The steepness of the progressions is influenced by the coal as well as the binding agent (Fig. 7). The F_{max} value of bituminous coal K22, for example, is not raised as much as that of steam coal K17 by the same quantities of binding agent. There is less difference in the fluidity values when coal and oil binding agents are added to bituminous coal than when these are added to steam coal.

The effect of the binding agents on the Roga Index and dilatation of steam coal K16 is illustrated in Fig. 5 by the examples of (only) 3 binding agents. In all three cases, the Roga Index increases constantly and almost linearly with the rising proportion of binding agent. The progressions of the lines vary considerably, however, the coal based binding agents again proving to be the most effective.

The dilatation values of the three blends reach a clear maximum when the proportion of binding agent is about 10%. This was also found to be the case for almost all the other blends which are not shown here.

If we take the criteria of a usable coking coal, i.e.

F_{max}	$>$	60 - 200 ddpm
dilatation	$>$	30%
Roga-Index	$>$	45

as a yardstick for assessing the suitability of binding agents, only carbo-pitch at a proportion of over 14% would meet these requirements in this case. The Roga Index is the critical parameter. In the case of steam coal K17, the above criteria can of course be obtained with smaller proportions of binding agents.

The fact that the maximum dilatation value of almost all blends occurs when the proportion of the binding agent reaches about 10% is linked with the special properties of the binding agent/coal blend. The dilatometer's load plunger is no longer raised by the swelling mass; instead, it keeps falling back because the cleavage products (mainly from the binding agent) easily penetrate to the outside in the form of gases or vapours through the highly fluid binding agent phase.

This means, however, that the dilatation values for blends containing a high percentage of binding agent are not completely accurate when obtained with an agitator dilatometer.

Apart from the above-mentioned parameter F_{max} serving as a fluidity criterion, the fluidity progression against temperature for various binding agent levels also provides insight into the interaction between the binding agent and the coal. As shown in Fig. 6, the F_{max} between 450 and 500°C increases with the proportion of binding agent almost linearly on a logarithmic graph, but at the same time the maximum clearly shifts to a lower temperature. At the same time, the plastic region (Gieseler region) indicated along the abscissa broadens as a result of the lowering of the softening temperature, whereas the re-solidification temperature remains constant within the limits of measuring accuracy.

With increasing percentage of binding agent, the fluidity curve levels off (14%) immediately after the softening stage at about 390°C, then reaches a maximum (15% and 20%) which goes beyond the measuring range of the Gieseler plastometer. Beyond that temperature, the fluidity again falls to a value corresponding to the undistasted plasticity progression and forms a second maximum within the Gieseler measuring range at about 470°C.

The first maximum temperature, however, in no way indicates the fluidity condition of a "combined molten mass": here soft superfluous binding agent acts as a lubricant between the vitrinite particles which are plasticised, but still for the main part solid. Depending on the quantity of superfluous binding agent, the fluidity progression indicates only a levelling-off as a result of the formation of a slip, or a maximum extending beyond the

Gieseler measuring range following the formation of an uninterrupted lubricant film between the slightly plastic or inert portions. The fluidity of the combined molten mass of steam coal-vitrinite and binding agent mesophase is expressed only by the true F_{max} value at 460°C, which accordingly continues to be increased by the larger quantity of binding agent.

This lubricating effect caused by superfluous binding agent is much stronger when the oil-based agent No. 7 is added; in this case, the F_{max} value increases regularly with the proportion of binding agent up to 21.5%, though only very slightly. When this "limit" is exceeded by only 1.1%, the system's fluidity shoots up and leaves the measuring range (over 50 000 grads/min). The transition temperature is roughly 300-330°C, which is similar to that for the binding agent carbo-pitch. The actual low F_{max} value therefore cannot be measured as a result of the "excessive" amount of bitumen added.

The high percentages of binding agent added in the studies are technically and economically irrelevant. For the modification of fluidity, the only binding agent proportions of interest are those between 5 and 10%.

In order to study the processes occurring in this blending range and in the temperature range between 390°C and 490°C, supplementary coking measurements are being carried out with the Ritter-Juranek small coking apparatus (4).

The main reason for the more positive effect of the coal-based binding agent is its "aromatic" character. The order of succession of binding agents listed according to increasing aromaticity (Table 2) is identical to that of the blends according to increasing F_{max} values for the same proportions of binding agent in the blends. The sole exception is bitumen No. 8, whose F_{max} values in the blends are too low compared with its aromaticity.

If we take the carbon distribution from the structural group analysis as a criterion we find a characteristic link between the F_{max} values and the percentages of carbon bonded in aliphatic methylene groups.

The coking behaviour of the coal/binding agent blends becomes more negative, the greater the proportion of C_{CH_2} in the binding agents; in this case, too, bitumen No 8 is out of line. The proportion of carbon in other aliphatic structural groups shows a similar but not so distinct tendency. In the case of aromatically bonded carbon, apart from carbon linked with hydrogen, carbon in the link points of aromatic rings in particular has a substantial influence on coking behaviour; this means, however, that a large proportion of highly condensed aromatic substances is required, as well as a high degree of aromaticity.

5.2. Effect of bitumen from the same type of coal

A question automatically posed when studying the correlation between binding agent properties and changes in the coking behaviour of coal/binding agent blends is that of the influence and effect on the coking properties of coal (of bitumen extracted from the same type of coal).

Although the results obtained with the SRC sample are good, this alone means very little since the material itself consists of hydrated constituents of coal which have been very highly modified by thermal means and have been formed from vitrinite, whereas ordinary bitumen mainly originates from the exinite group of macerals.

The influence of original coal bitumen can be studied

- on the basis of blends of original coal and isolated coal bitumen (i.e. by adding various quantities of bitumen)
- by partial or total extraction of coal bitumen (i.e. by varied degrees of bitumen removal).

The latter method was applied in numerous studies in the past described by various authors (5). We shall see below, however, that this process is bound to lead to the wrong conclusions.

5.2.1 The addition of bitumen

When bitumen from the same type of coal is added (0-15%) to steam coal K17 and bituminous coal K22, their maximum fluidity is increased in the same linear manner as that of other binding agents on a logarithmic graph (Fig. 7). The increase in maximum fluidity is similar to that of the other coal-based binding agents; it is not as sudden as that of carbo-pitch blends but much higher than oil bitumen No 3. The softening temperatures of the blend are linearly lower and consequently the plastic region is significantly broadened. The resolidification temperatures, on the other hand, hardly vary and are only slightly higher than those of original coal.

The changes in the softening and resolidifivation temperatures shown by the dilatometer are similar to those shown by the Gieseler Plastometer; however, the resolidification temperature falls slightly but significantly.

By adding 2%, 4% and 6% own bitumen to K22, the dilatation is increased sharply (Fig. 8). By contrast, contraction varies only within the limits of reproducibility; the temperature at maximum contraction (commencement of swelling), however falls sharply like the softening temperatures. The comparably high Roga Index of the basic coal is improved only slightly from 65 to 69 by the addition of bitumen.

In addition to experiments on the addition of complete own bitumen, the proportion of bitumen which is insoluble in petroleum ether (solid bitumen) and the portion of bitumen which is soluble in petroleum ether (oil bitumen) were blended with the various types of coal (Fig. 9).

When oil bitumen or solid bitumen is blended with steam coal K17, the latter's softening temperature is lowered by 55 to 60K, and consequently the plastic range of 59K is practically doubled to 116 or 117K (the resolidification temperature remaining constant).

The results of the blending of oil or solid bitumen with bituminous coal K22 are similar, but in this case the softening temperature falls by only 30K; this is partly because the softening temperature is already low and partly because of the small amount of bitumen added (4%). These are also the reasons for the only slight broadening of the plastic range from 84K to 101K.

The plastic region is therefore broadered just as effectively with the addition of solid bitumen as with the respective oil bitumen, but only in cases where the solid bitumen originates from the same type of coal. If, on the other hand, bituminous coal K22 is mixed with a bitumen from a different coal type, e.g. from gas-flame coal K37 (FB_{37}) the softening temperature does not fall as much as for an oil bitumen; the plastic range is also reduced.

The addition of oil bitumen from other types of coal, on the other hand, causes a fall in the softening temperature and broadening of the plastic range similar to that caused by oil bitumen from the same type of coal.

The contraction values vary only within the scatter area. The commencement of swelling is lowered by about 20K in all blends (oil and solid bitumen blended with K17 and K22). When solid bitumens from other types of coal (FB_{87}) are added, on the other hand, the commencement of swelling is reduced by only 10K.

Dilatation is increased by the two types of oil and solid bitumen in the case of K17 and K22. For K22, the effect of oil and solid bitumen is the same; for K17, solid bitumen increases dilatation and consequently raises the G value much more sharply than oil bitumen.

If solid or oil bitumen from a different type of coal (FB_{37}, $ÖB_{37}$) are added in place of those from the same type of coal, dilatation increases but not as strongly as in a blend with part of the coal's own bitumen.

When FB_{37} and $ÖB_{37}$ are added to bituminous coal 22, dilatation is affected in more or less the same way; in the case of steam coal K17, on the other hand, solid bitumen from a different coal causes greater dilatation than oil bitumen from a different coal.

The Roga Index, serving as a yardstick for the bonding capacity for inert substances, varies according to the use of oil or solid bitumen for both bituminous coal K22 and steam coal K17.

The effect of adding solid bitumen is basically positive, i.e. the Roga Index is increased every time, and in some cases very substantially. The addition of oil bitumen, on the other hand, causes a reduction in the Roga Index in every case (Fig. 10), irrespective of the origin of the oil bitumen, i.e. it does not matter whether it is derived from the same type of coal or from a gasflame coal ($ÖB_{37}$).

To sum up, the findings are as follows:

when blended with steam coal,

- solid bitumen causes greater dilatation than oil bitumen,
- solid bitumen improves the Roga Index, whereas oil bitumen reduces it.

5.2.2 Bitumen extraction

It has long been known that coal whose bitumen has been removed by extraction has only a very negligeable caking or coking capacity (5). The extraction has been carried out mostly with benzene at high temperatures of 260-280°C in autoclaves at a service pressure of about 50 bar.

A quantitative link between the extract yield and coking capacity of the tested coal types was established but it cannot be formulated in exact mathematical terms. The link between the quantity of extracted bitumen and the coking properties of the residue coal (extraction residue) is even more difficult to formulate accurately. In most cases, the loss of caking and coking capacity was attributed to the removal of bitumen, which was the main material with these properties. This explanation is only partly correct; a more precise examination of the "pressure extraction" stage of the process shows in fact that the coking properties are altered fundamentally without any bitumen extraction taking place when the coal is subjected to pressure decomposition.

The dilatation of bituminous coal K22, for example, is reduced from +27% to -22%, i.e. by 49% in absolute terms, as a result of this three-hour decomposition of coal with benzene at 280°C (described as "manipulation" below (Fig. 11). At the same time, the commencement of swelling in the dilatometer is postponed until the temperature is 10° higher without any notable change in the softening temperature; the extent of the plastic region, however, remains unchanged.

The caking capacity is also reduced considerably as a result of this manipulation: e.g. the Roga Index falls from 65 (for the original coal) to 24 (for the manipulated coal), i.e. the coal's capacity for bonding inert substances is reduced substantially as a result of the pressure decompositon process.

This change is presumed to be caused by a partial destruction of the "colloidal structure" as a result of a loosening of the bitumen parts of the coal matrix and possibly a concentration of bitumen on the surface of the particles. The following observations give weight to this assumption:

- the volatile constituents of manipulated coal (MK) are 1-5% (relatively) higher than for original coal (OK), depending on the decomposition temperature selected (160-305°C)
- a purely thermal treatment in an argon atmosphere or decomposition by water cause only slight changes in the dilatometer parameters.

Dilatation is lowered by only 6% from -22% to a pure contraction of 28% by the actual extraction of bitumen, i.e. the removal of the benzene-soluble parts from the manipulated coal. The softening temperature of this residue coal (RK) is 40K and the commencement of swelling 20K higher than for the original coal (Fig. 11).

The restoration of the extracted bitumen to the residue coal (RK + 5.2% GB) does not restore the original coal's coking properties at all. In the case of the K22 coal, only the dilatometer parameters of the manipulated coal were restored more or less to their original level. Similar results were obtained for a series of other types of coal examined (3).

A notable aspect of these findings is the fact that bitumen extraction and restoration have an effect on the softening temperature, plastic region, commencement of swelling and dilatation level, whereas the pressure decomposition as such only alters dilatation and the commencement of swelling, the softening temperature and plastic region, on the other hand, remaining unchanged.

The caking capacity, expressed by the Roga Index, is very much reduced by bitumen extraction and not fully restored by the replacement of bitumen, as shown by the following figures:

		Roga Index
Original coal	OK 22	65
Manipulated coal	MK 22	24
Residue coal	RK 22	14
Residue coal +5.2% bitumen		21

Bibliography

1. Beck, K.-G. und Meckel, J.
Glückauf 115 (1979), S. 979-983

2. - Miyazu, T.; Okuyama, Y.; Fukuyama, T. et. al.
International Iron and Steel Congress, Düsseldorf 1974
- Miyazu, T.; Matsubara, K. und Morotomi, H.
Ironmaking Proceedings Vol. 37, Chicago 1978

3. Abel, O.; Gao, J.; Oelert, H.-H.
Round-Table-Konferenz der KEG am 24./25. Nov. 1980, Brüssel

4. Ritter, H. und Juranek, G.
Brennstoff-Chemie 41 (1960), S. 170-176

5. - Fischer, F. et al.
Brennstoff-Chemie 5 (1924), S. 299
6 (1925), S. 33
6 (1925), S. 349
- Wheeler, R.V. et al.
J. Chem. Soc. 99 (1911), S. 649
103 (1913), S. 1704
107 (1915), S. 1318
- Bone, W.A. et al.
Proc.Roy. Soc. A105 (1924), S. 608
A120 (1928), S. 523

Samples		Elementary Composition			Molar mass	Relative carbon distribution											
						aliphatic			stable aromatic compounds					aromatic			
									final			transitional					
		C	H	H/C	M	CH_3	CH_2	CH	CH_3	CH_2	CH	CH_2	CH	CH	$C_{subst.}$	$C_{cond.}$	f_a
Carbopitch		92,7	5,1	0,66	351	4,5	4,9	0	0	24,6	-	-	-	32,6	6,2	47,3	0,86
SRC		85,4	6,9	0,97	-	-	-	-	-	-	-	-	-	-	-	-	-
K 17	Own bitumen	88,8	6,5	0,88	314	2,2	4,2	2,8	6,5	1,7	1,1	0,8	0,5	43,9	14,0	22,3	0,80
	Oil bitumen	89,7	7,2	0,95	286	3,8	7,8	5,2	4,9	4,2	2,8	0,5	0,3	35,9	15,3	19,5	0,71
	Solid bitumen	87,5	5,7	0,77	453	1,8	3,7	2,5	4,9	3,5	2,4	0,7	0,4	36,5	15,7	28,0	0,80
K 22	Own bitumen	87,2	6,6	0,90	411	2,8	7,5	5,0	7,8	1,6	1,1	0,6	0,4	32,4	14,1	26,9	0,73
	Oil bitumen	88,9	7,4	1,00	313	4,4	10,1	6,7	6,5	3,5	2,3	0	0	31,0	13,9	21,6	0,67
	Solid bitumen	86,1	5,9	0,82	699	1,5	5,5	3,7	5,5	3,8	2,5	1,1	0,7	33,8	18,5	23,5	0,76
Bit. 3		82,9	8,5	1,21	843	6,6	15,5	10,4	8,7	4,0	2,7	2,4	1,6	14,9	25,7	7,4	0,48
Bit. 4		83,7	9,4	1,33	1091	7,2	19,9	13,3	11,6	1,1	0,7	2,7	1,8	11,5	25,0	5,2	0,42
Bit. 7		83,2	10,2	1,47	1422	9,4	24,1	16,0	10,2	0,5	0,4	3,0	2,0	12,0	23,9	0	0,34
Bit. 8		83,9	8,7	1,23	964	9,3	15,7	10,5	8,8	2,6	1,7	1,6	1,1	14,1	20,3	14,4	0,49

Table 2: Elementary composition and relative carbon distribution of the binding agents

Samples	K 16	K 17	K 22
Water content	0,92	0,63	0,64
Ash (wf)	5,72	2,3	2,1
Volatile constituents (waf) %	16,3	17,5	22,3
Swelling Index	-	2 1/2	8 1/2
Roga - Index	14	41	65
Dilatometer parameters:			
Softening temperature °C	458	439	397
Resolidification temp. °C	499	498	481
Extent plastic region K	41	59	84
Contraction %	12	25	28
Dilatation %	-4	6	27
Coking capacity G	0,921	0,962	0,998
Plastometer parameters (Gieseler):			
Softening temperature °C	475	456	430
Temperature at F_{max} °C	475	468	463
Resolidification temp. °C	475	476	484
Extent plastic region K	1	20	54
Maximum fluidity (grads/min)	1	2	23

Table 1: Analytical data of coal types

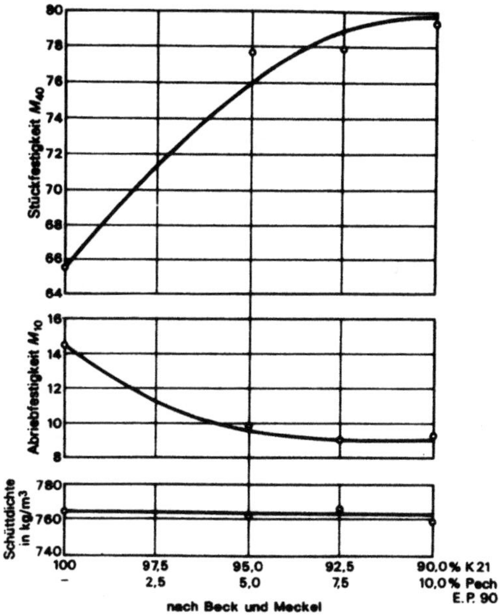

Fig. 1

Schüttdichte = Bulk density
Abriebfestigkeit = Abrasive resistance
Stückfestigkeit = Impact resistance
Pech = pitch

after Beck and Meckel

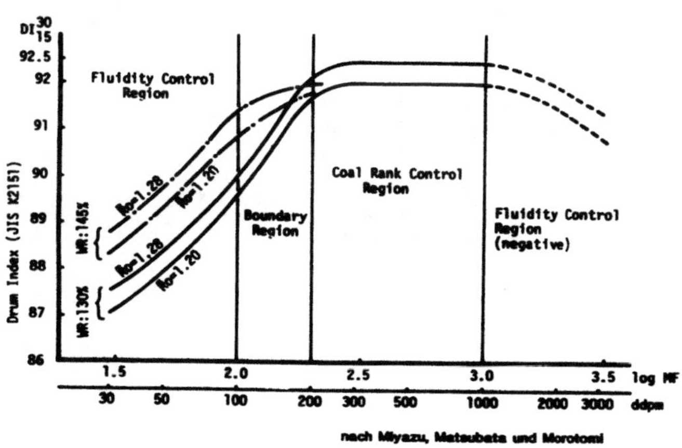

Fig. 2

after Miyazu, Matsubata and Morotomi

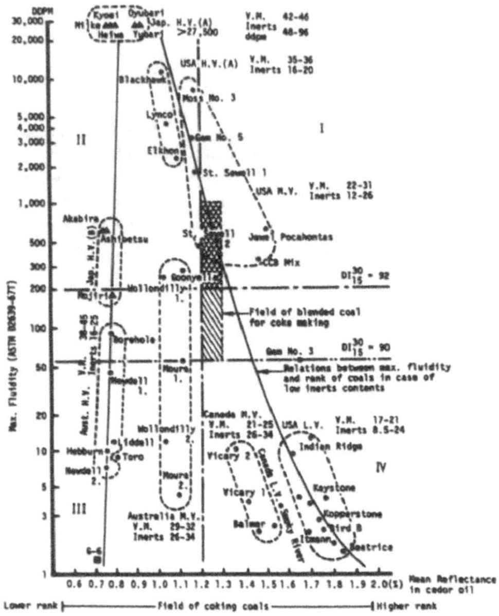

nach Miyazu, Matsubata und Morotomi

Fig. 3

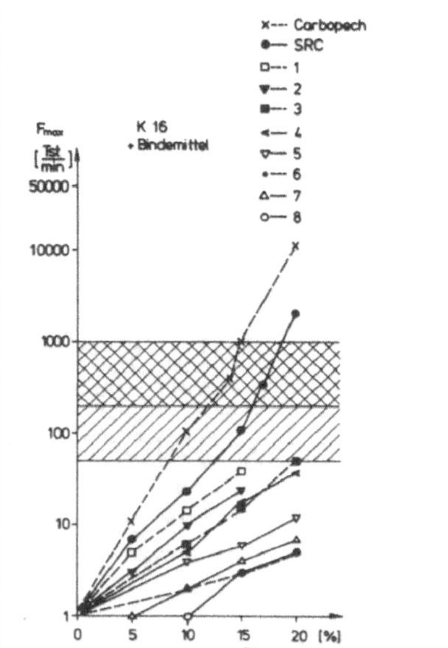

Einfluß von Qualität und Mengenanteil des Bindemittels auf die Maximale Fluidität (F_{max}) einer K16

Fig. 4

Influence of quality and quantitative proportion of the binding agent on the maximum fluidity (Fmax) of a K16 coal

Carbopech	**=**	**Carbopitch**
Bindemittel	**=**	**Binding agent**
Bindemittelanteil	**=**	**Binding agent proportion**

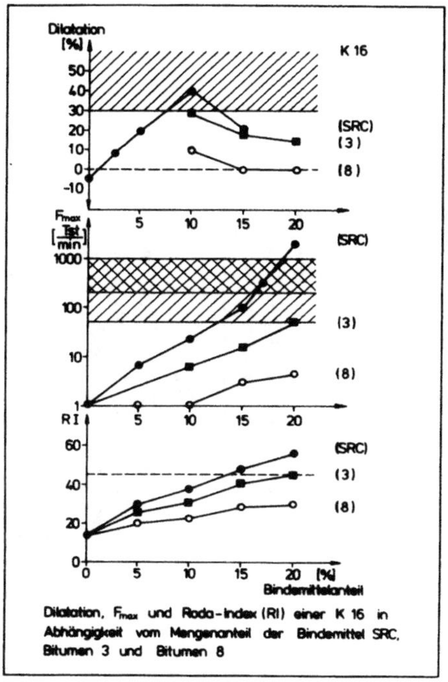

Fig. 5

Dilatation, Fmax and Roga Index (RI) of a K16 coal as a function of the quantitative proportion of binding agents SRC, Bitumen 3 and Bitumen 8

Bindemittelanteil = Binding agent proportion

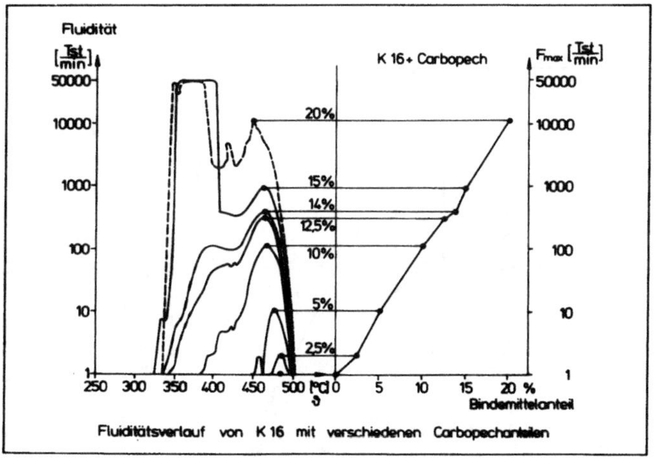

Fig. 6

Fluidity progression of K16 with varying proportions of carbo-pitch

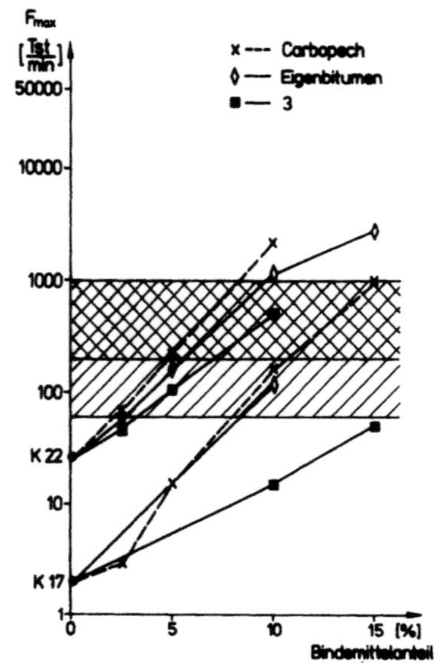

Einfluß von Carbopech, Eigenbitumen und Petropech 3 auf die Maximale Fluidität (F_{max}) einer K 17 und K 22

Fig. 7

Influence of carbo-pitch, own bitumen and petro-pitch on the maximum fluidity Fmax of K17 and K22 coal ranks

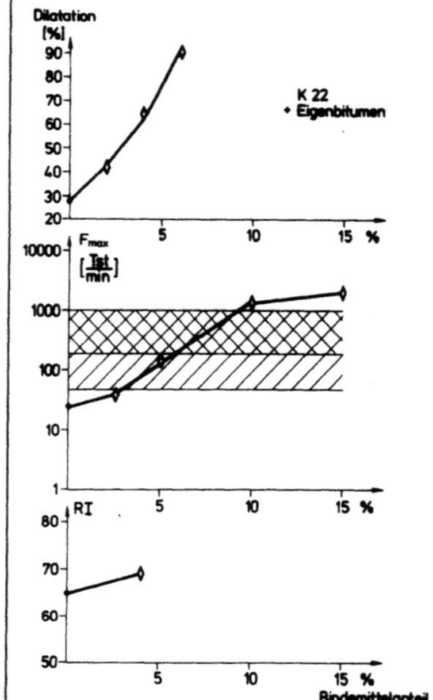

Veränderung von Dilatation, F_{max} und Roga-Index (RI) bei Zugabe von Eigenbitumen zu einer K 22

Fig. 8

Alteration of Dilatation Fmax and Roga Index (RI) by addition of own bitumen to a K22 coal

Carbopech	=	Carbopitch
Eigenbitumen	=	Ownbitumen
Bindemittelanteil	=	Binding agent

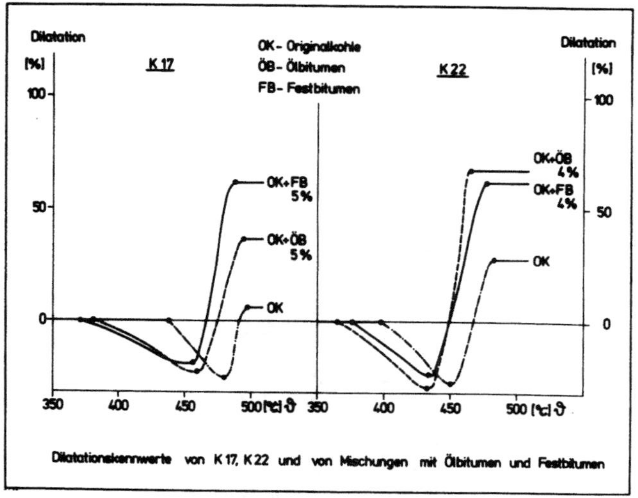

Fig. 9

Dilatation parameters of K17, K22 and blends containing oil bitumen and solid bitumen

OK = Original coal
OB = Oilbitumen
FB = Solidbitumen

Proben	Roga - Index absolut	Roga - Index relativ % bezogen auf Originalkohle	Dilatation absolut	Dilatation relativ % bezogen auf Originalkohle
K 17	41	100	6	100
+ Ölbitumen 5 %	22	54	35	583
+ Festbitumen 5 %	53	129	61	1016
+ $Ölbitumen_{37}$ 5 %	34	83	24	400
+ $Festbitumen_{37}$ 5 %	53	129	34	567

Einfluß von Ölbitumen und Festbitumen auf Roga - Index und Dilatation einer K 17

Fig. 10

Influence of oil bitumen and solid bitumen on the Roga Index and dilatation of a K17 coal

Proben = Samples
relativ % bezogen auf Originalkohle = relative % in relation to original coal
Ölbitumen = oil bitumen
Festbitumen = solid bitumen

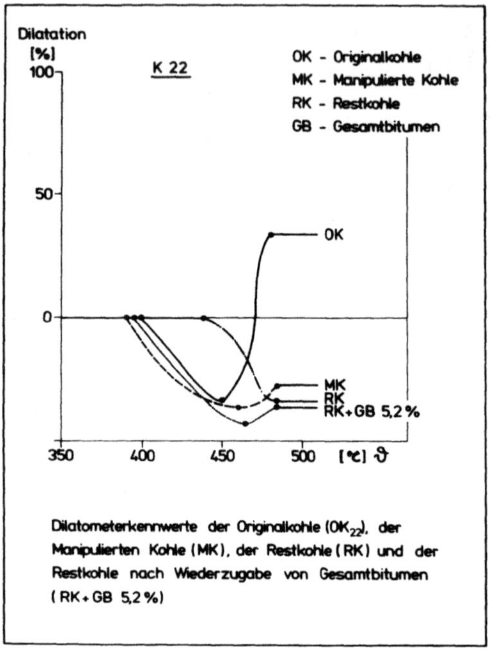

Fig. 11

Dilatometer parameters of original coal (OK_{22}), manipulated coal (MK), residue coal (RK) and residue coal after total bitumen restoration

OK = Original coal
MK = Manipulated coal
RK = Residue coal
GB = Total bitumen

HIGH TEMPERATURE STABILITY OF COKE IN RELATION TO ITS BULK PERMEABILITY

J. GOLECZKA and J. TUCKER

Coal Research Establishment, National Coal Board

Summary

In the blast furnace, the initial size grading of the feed coke and subsequent changes in its lump size distribution can markedly affect the permeability of the stock column and hence furnace performance. A realistic assessment of the role of coke as a burden spacer requires measurements of its mechanical size stability at conditions simulating the chemical environment and thermal stresses of the smelting process, using bulk samples of the coke as charged.

The thermo-mechanical size stability of a wide range of cokes has been evaluated by a test based on the rapid reheating of 70 kg samples to 1400°C in an inert atmosphere, followed by cooling to ambient temperature and micum-drum testing. The results have shown that the bulk thermal stability of coke depends on the rank of the parent coals and blend composition, the oven charge preparation and carbonisation conditions, and the lump size of the cokes produced. In general, the thermal stability of coke cannot be predicted from tests at ambient temperature alone.

Complementary measurements of the bulk reactivity and post-reaction strength of coke have been carried out by reacting, initially, 25 kg samples with carbon dioxide at 1000°C to a weight loss of up to 60%.

1. INTRODUCTION

The increasing scale and intensity of modern blast-furnace operation, combined with a simultaneous reduction in coke rate, have led to more stringent coke quality requirements with a greater emphasis on the coke's role as a burden spacer[1,2]. Accordingly, coke quality data limited to chemical composition, initial size grading and ambient size stability are insufficient to predict the coke's behaviour in use. A realistic assessment of the likely performance of coke should also take into consideration its resistance to degradation in the chemical and thermal environment of the blast furnace. A test is, therefore, needed which simulates as far as possible the mechanical, chemical and thermal conditions of the smelting process and measures the resultant changes in size distribution of the coke, and hence in its bulk permeability to gas flow, using representative samples of the feed coke as charged to the furnace. Moreover, new developments in coke production make it necessary to reassess coke testing so as to guide the coke producer in making the best use of available coals.

The mode of mechanical coke breakage at ambient temperature has been extensively characterised and related to changes in the coke's bulk permeability[3]. In contrast, the causes of deterioration in the hot strength and size of bulk coke arising from its partial gasification with carbon dioxide and reheating to high temperature are relatively less explored. The present paper describes the work carried out at the Coal Research Establishment (CRE) concerning the bulk thermal stability of a wide range of cokes, i.e. their resistance to mechanical degradation following an exposure to temperatures well above those of carbonisation. Some results are also included of a separate, complementary study of the bulk chemical stability of coke.

2. BULK THERMAL STABILITY OF COKE

2.1 Method of measurement

The development and details of the CRE bulk thermal stability test have been described in previous papers[4-7]. The basic aim has been that coke of the size grading fed to blast furnaces should be heated to temperatures and at rates approaching those encountered in the bosh but that it should not undergo any significant chemical reaction with its environment. Accordingly, the method developed involves the rapid reheating of 70 kg of coke to about 1400°C in an oven-furnace purged with nitrogen. After a heat soak at 1400°C followed by a slow cooling to ambient temperature, the coke is tested mechanically in a half-micum drum. The irreversible nature of the thermal effects renders the mechanical testing of coke in the hot state unnecessary.

2.2 Affected coke properties

An example of average changes in coke properties upon reheating to 1400°C is shown in Table 1. With a net loss in weight of 5%, there were contributory reductions in volatile matter, ash and sulphur contents. In addition to a densification of the coke substance, coke lump size was also reduced and mechanical testing in the micum drum markedly increased the size difference between the original and reheated coke samples. Clearly, the reheating had weakened the cokes and substantially lowered their resistance to volumetric 'coarse' breakage (harmonic mean size of plus 10 mm debris, S_h^{100}) and their resistance to surface degradation (M'_{10} and I_{10} indices).

Using a previously established relationship[(3)], Figure 1, the aerodynamic properties of cokes can be predicted from data on harmonic mean size and uniformity of size distribution before and after reheating and various stages of treatment in the micum drum. It is evident that the direct effect of changes in the harmonic mean size of cokes with rising severity of degradation is reflected in progressively greater increases in their bulk resistance to gas flow.

2.3 Effect of initial coke lump size

Tests on closely sized coke fractions permitted a comparison of the thermal stabilities of different cokes to be made on a common basis independent of their initial size distributions. Figure 2 gives details of the results obtained with coke samples W1-W4 representing four different batches of a commercial blast-furnace coke sampled over separate production periods. Although the mechanical properties of all four coke consignments were similar at ambient temperature in terms of harmonic mean size (Figure 2a) it is evident that reheating the coke samples to 1400°C exposed marked differences in their resistance to thermo-mechanical breakdown. This is emphasised still further (Figure 2b) by their relative thermal stability (RTS) index, i.e. the percentage ratio of the harmonic mean size of reheated to ambient coke debris after micum testing.

The results, therefore, confirmed that prediction of the thermal behaviour of coke was not possible from tests at ambient temperature alone. They also indicated that the relative resistance to thermal breakage (RTS index) decreased with increasing coke lump size. It is noteworthy that both the harmonic mean size and relative thermal stability values obtained from tests on individual size fractions were additive since the corresponding values for the representative plus 20 mm coke samples fitted satisfactorily on the lines plotted.

2.4 Effect of oven-charge carbonisation conditions

The reasons for the variation in thermal stability parameters of the four coke batches (W1-W4) were not apparent from plant data on parent coal properties, oven charge preparation or carbonisation conditions. However, laboratory measurements in an ultra-high temperature (UHT) dilatometer showed variations in the dimensional changes of these cokes upon gradual heating to 1500°C. Thus, after an initial uniform expansion, the onset and extent of their subsequent contractions were different. The corresponding blast-furnace tuyere cokes expanded over a longer temperature range because of the more severe heat treatment received in the blast furnace. Figure 3 shows that these trends were reflected in the corresponding coefficients of linear change; their maximum values (α) of (tertiary) contraction being indicative of the magnitude of thermal stresses likely to occur within the cokes during reheating. It was not surprising, therefore, that the α values of coke samples W1-W4 were consistent with the results of thermal stability tests shown in Figure 2, i.e. the lower the maximum coefficient of (tertiary) contraction the better the thermal stability of the coke. Both parameters appeared to have been influenced by the coke finishing conditions in the commercial ovens.

2.4.1 Finishing temperature

Tests on cokes produced at a constant (standard) flue temperature in a 500 kg pilot oven confirmed the importance of adequate finishing in the oven on the cokes' thermo-mechanical size stability. Figure 4 indicates that although the ambient size stability of the cokes was virtually unaffected by increasing the final charge-centre temperature from 900°C to nearly 1150°C, the thermal stability (residual harmonic mean size) increased progressively with rising temperature and thermal uniformity of the final stage of carbonisation. It can be seen that the latter trend was accompanied by decreasing values of the cokes' peak contraction coefficient (α).

2.4.2 Additional isothermal holding

A clear distinction between the effects of coke finishing temperature and additional isothermal holding of the charge in the oven is shown in Figure 5, based on carbonisation tests in an electrically-heated 300 kg pilot oven. It is evident that the thermo-mechanical size stability of the cokes made at 900°C and 1000°C was enhanced by holding them at their respective final temperatures for an additional period of 3 hours. However, this did not apply to coke made at 1100°C, indicating that the oven charge had been fully carbonised upon reaching this temperature.

The corresponding coefficients of linear change derived from UHT dilatometer measurements and plotted in Figure 6 had maximum values (α) of coke (tertiary) contraction which were clearly dependent, with regard to both magnitude and temperature of occurrence, on coke finishing temperature and additional isothermal holding in the oven. It is also obvious from this figure (and from Figure 3) that the temperature of the onset of coke contraction, the maximum 'rate' of contraction (α) and its temperature of occurrence were closely inter-related. Further results embracing a wider range of cokes confirmed the strong tendency for the coefficient (α) values, and hence for the thermal stressing of the coke matrix during reheating, to decrease with increasing temperature of occurrence and, consequently, with rising thermal level and uniformity of carbonisation. In this context, it is debatable to what extent the additional thermal treatment that coke receives in the blast-furnace stack, prior to entering the bosh zone of rapid temperature rise above 1100°C, can thermally upgrade a poorly-carbonised coke and thus improve its thermo-mechanical stability.

2.4.3 Carbonising rate

The influence of charge carbonising rate on coke quality was examined in the 500 kg pilot oven by altering the oven-flue temperature whilst keeping the charge finishing temperature constant at 1020-1040°C. Figure 7 shows that, as expected, the extension of carbonising time from 15-16 hours to about 20 hours increased the ambient size of the plus 20 mm coke. Moreover, this was reflected in the lump size of the mechanically tested products both before and after reheating to 1400°C.

2.5 Effect of parent blend properties

The rank effect of a wide range of coals of adequate caking capacity (total dilation $\geq$ 45-50%) on the thermo-mechanical size stability of resultant cokes is presented in Figure 8 as a function of oven-charge volatile matter content. It can be seen that the extent of deterioration in the harmonic mean size of the individual cokes arising from reheating was not uniform. By contrast, the residual size stability of the reheated cokes decreased in an approximately linear manner as the volatile matter content of the oven charges increased from 18% to 37% d.a.f. As shown elsewhere[5,7], there was a corresponding fall in the cokes' bulk permeability to gas flow which suggested that a more stringent blend formulation might be necessary than hitherto accepted on the basis of ambient mechanical tests alone.

In separate investigations, small additions of finely crushed coke-oven breeze to coal charges were shown to be beneficial with regard to overall coke lump size and thermo-mechanical stability, but at a cost of increased sensitivity to surface degradation[6,7].

2.6 Effect of oven charge preparation

The production of coke of high and uniform quality requires not only a rational selection of suitable coals and additives for blending but also a close control of blend preparation procedures[8].

2.6.1 Crushing

It is known that the degree of coal crushing may influence the mechanical stability of the coke product, particularly if the mineral matter present in the individual blend components is not finely disseminated. Figure 9 indicates that, in the case of blend A' (V.M. 32.3% d.a.f.) coarse crushing of the oven charge was undesirable because of reductions in ambient and thermal lump size stabilities of the coke, accompanied by a marked deterioration in its resistance to surface breakage. Similarly the detrimental effect of coarse mineral matter was clearly demonstrated by a small inclusion of washery discard in blend B (V.M. 27.6% d.a.f.).

2.6.2 Densification

Increasing the bulk density of coke-oven charges is known to improve the apparent density and surface cohesion of the coke product and extend the range of coals that can be used for coke-making. Figure 10 shows that in the case of blend C (V.M. 33.4%, 35% ICC 721 coal) fluxed with 3% of coal-tar pitch, charge densification by either drying or partial briquetting resulted in smaller (plus 20 mm) cokes with progressively better resistance to surface degradation. After reheating to 1400°C and mechanical testing, the harmonic mean sizes of the cokes were equalised and the improvements in coke-surface cohesion were fully retained.

2.6.3 Preheating

As illustrated in Figure 11 with three different blends, the improved resistance of the cokes to surface degradation following charge preheating and pipe-line charging was maintained after exposing the cokes to higher temperatures than those at which they were produced. However, the beneficial effect of charge preheating was not manifest in the size stability of these cokes either in the ambient test or after reheating.

3. BULK CHEMICAL STABILITY OF COKE

As mentioned earlier, the chemical aspects of coke behaviour in the blast-furnace stack are being assessed in a separate CRE study, complementary to that on thermal stability. There are conflicting arguments concerning the merits of various coke reactivity tests. However, the presence of alkali compounds in the blast furnace means that the extent and mode of coke gasification with carbon dioxide (CO_2) may be more dependent on furnace operating conditions, and consequent retention of alkalis, than on coke properties. Recent sampling exercises indicate that about 20% to 30% of the feed coke is gasified by the carbon solution loss reaction[9]. This suggests that, for a direct comparison between cokes, measurements of their residual strength after reaction to a given weight loss (say 20% or 30%) may be preferable, as a measure of the cokes' potential performance in the blast furnace, to an assessment simply of their reactivity.

It is well known that the apparent density and mechanical quality indices of coke lumps vary according to the original position in the oven during carbonisation. The example shown in Figure 12 confirms that the heterogeneous nature of coke is also reflected in its reactivity to CO_2 thus emphasising the necessity of testing representative samples of the bulk coke as produced. A further reason for bulk reactivity tests is that measurements can be made of both modes of coke degradation (volumetric and surface breakage) using the micum drum after cooling the coke residue to ambient temperature.

Accordingly, a method of reacting uniformly 25 kg samples of coke in a static bed, initially with CO_2 at 1000°C, has been developed using specially designed Refrax (silicon carbide) reaction vessels. Figure 13 shows the post-reaction strength indices of two cokes after a 20% weight loss, as a function of original lump size between 20 and 80 mm. It is evident that the resistance of the gasified cokes to surface degradation was markedly reduced; the extent of this deterioration could not be predicted from ambient mechanical tests alone. There was also a parallel drop in coke lump size. Although significant differences were noted between the reactivities of the two cokes tested, the corresponding rates of reaction were virtually independent of original lump size. Figure 14 confirms that the effect of extended coke gasification up to 60% weight loss was manifest in further progressive reductions in its resistance to surface degradation and in harmonic mean size.

The combined effects of both chemical and thermal influences on the size degradation of coke, and consequently on its residual bulk permeability, are understood from other investigations(10) to be approximately additive. The results of CRE tests shown in Figure 15 indicate the respective actions of the individual and combined treatments on coke quality parameters. Thus, direct reheating to 1400°C (T) provides a severe test of the thermo-mechanical lump-size stability of coke whereas the superimposed bulk reactivity test (R+T) emphasises the reduction in the coke's post-reaction resistance to surface degradation. The stabilising effect on lump size of an equivalent thermal holding of the coke in an inert atmosphere prior to reheating (T') is shown by the intermediate 'blank' test.

4. CONCLUSION

It is evident from the CRE studies that both blend composition and carbonisation variables affect the high temperature stability of the coke product. Further work is necessary in this field to derive meaningful indices of coke quality relevant to its performance in use.

5. REFERENCES

1. Higuchi, M. and others. Metals Society Conference on Coal, Coke and the Blast Furnace, Middlesbrough, June 1977 (Paper 2).
2. Winzer, G. and Engel, K. (As ref. 1, Paper 3).
3. Gibson, J. and Goleczka, J. CEC Information Symposium on Coke Oven Technicques, Luxembourg, May 1975. Part 1, p.275.
4. Goleczka, J. and Roberts, T. L. (As ref. 1, Paper 8).
5. Goleczka, J. and Roberts, T. L. ECSC Round Table Meeting on Coke Oven Techniques, Luxembourg, October 1978, p.221.
6. Goleczka, J. and Tucker, J. ECSC Round Table Meeting on Coking Plants and Coke, Luxembourg, October 1980.
7. Final report on ECSC Research Project No. 7220-EB/811, March, 1981.
8. Gregory, D. H. and Peirce, T. J. (As ref. 1, Paper 11).
9. Adair, R. R. and others. (As ref. 5) p. 186.
10. Miyazu, T. and others. (As ref. 1, Paper 7).

TABLE 1.

EXAMPLE OF AVERAGE CHANGES IN COKE PROPERTIES UPON REHEATING TO 1400 °C.

COMMERCIAL COKE: BATCHES W1 - W4
STANDARD INDICES: M_{40} = 81 - 82, M_{10} = 6

STATE OF COKE		ORIGINAL	REHEATED
WEIGHT LOSS ON REHEATING	%	-	5
VOLATILE MATTER CONTENT (D.A.F.)	%	0·8	0·2
ASH CONTENT (D.B.)	%	6·9	6·5
SULPHUR CONTENT (D.B.)	%	1·1	1·0
APPARENT DENSITY	gcm^{-3}	0·84	0·85
TRUE DENSITY	gcm^{-3}	1·94	1·99
POROSITY	%	56·8	57·2
HARMONIC MEAN SIZE, S_h	mm	60·1	57·4
MICUM DRUM TESTS ON +20mm COKE: HARMONIC MEAN SIZE, S_h^{100} OF +10mm DEBRIS AFTER 100 REV.	mm	49·1	40·4
SURFACE DEGRADATION INDICES: M'_{10} (%WT. - 10mm, 100 REV)		5·9	8·2
I_{10} (%WT. - 10mm, 500 REV.)		20·6	23·5

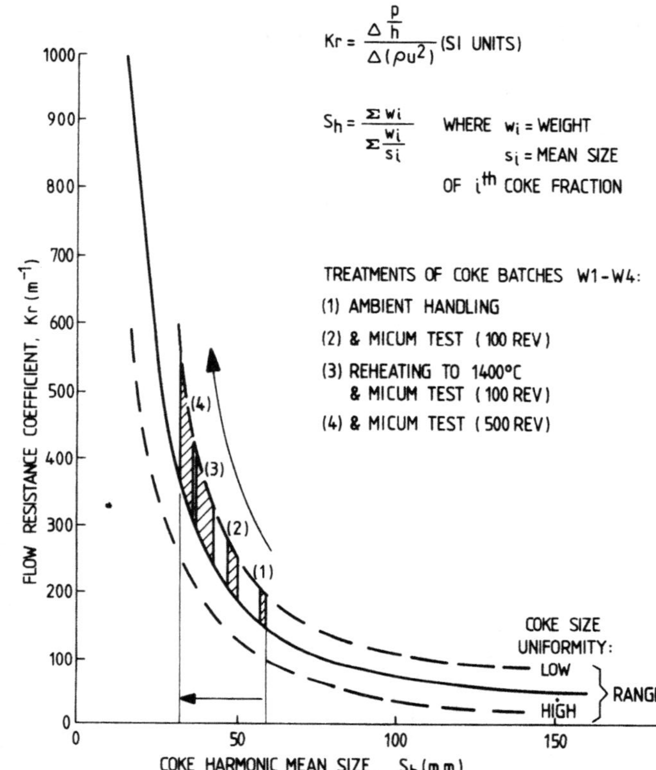

FIGURE 1. EFFECT OF COKE SIZE DEGRADATION ON BULK RESISTANCE TO GAS FLOW.

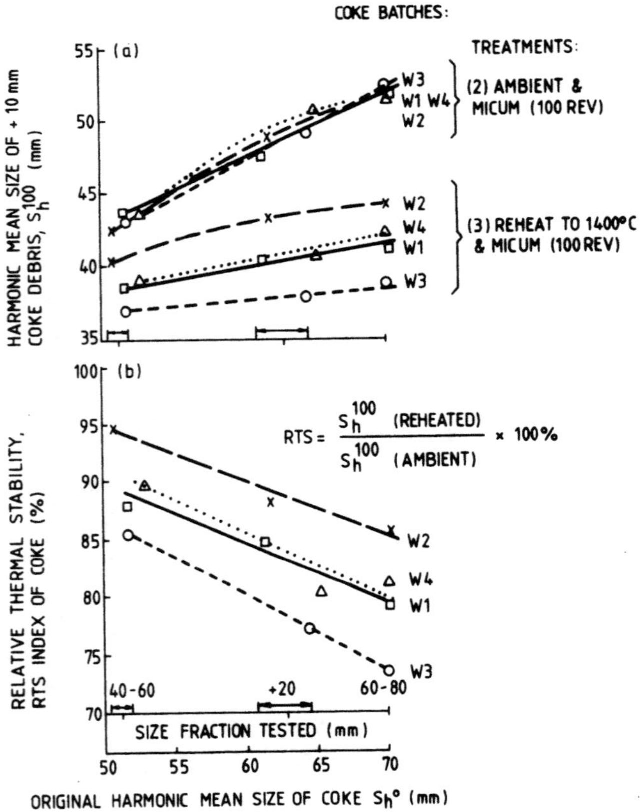

FIGURE 2. THERMO-MECHANICAL SIZE STABILITY INDICES OF DIFFERENT BATCHES OF A BLAST-FURNACE COKE.

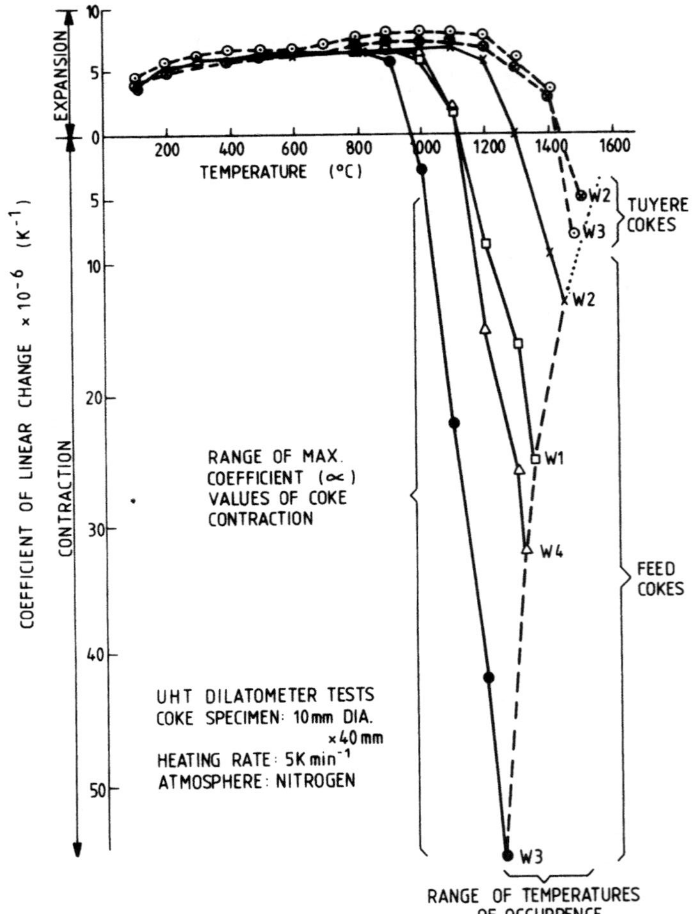

FIGURE 3. COEFFICIENTS OF LINEAR CHANGE OF DIFFERENT BATCHES OF A BLAST-FURNACE COKE.

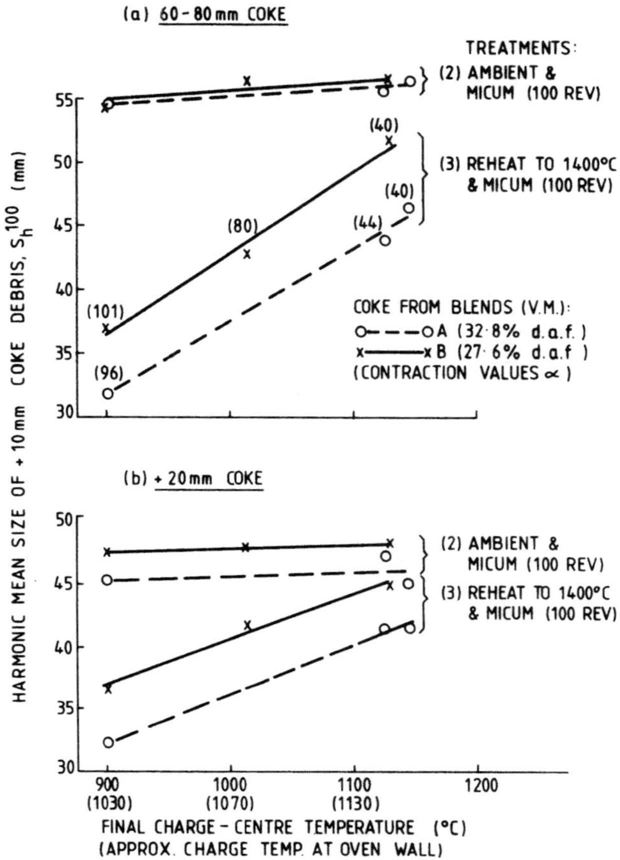

FIGURE 4. EFFECT OF CHARGE FINISHING TEMPERATURE IN OVEN ON THERMO-MECHANICAL SIZE STABILITY OF COKE.

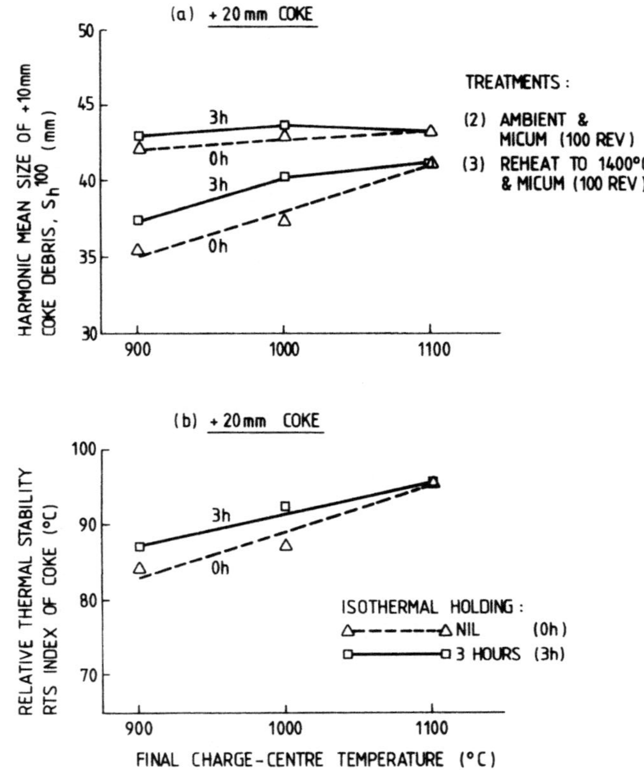

FIGURE 5. EFFECT OF ISOTHERMAL HOLDING OF OVEN CHARGES AT DIFFERENT FINISHING TEMPERATURES ON THERMO-MECHANICAL SIZE STABILITY INDICES OF COKE.

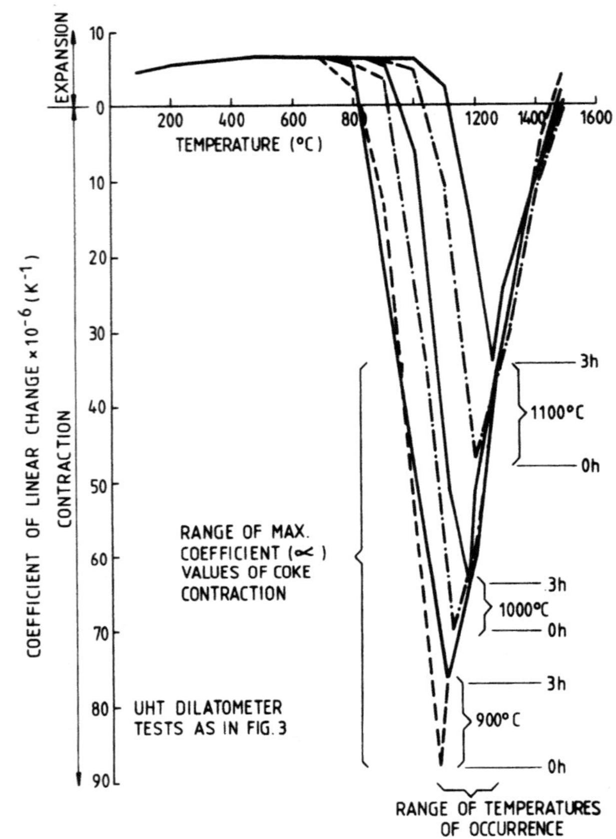

FIGURE 6. EFFECT OF ISOTHERMAL HOLDING OF OVEN CHARGES AT DIFFERENT FINISHING TEMPERATURES ON COEFFICIENTS OF LINEAR CHANGE OF COKE.

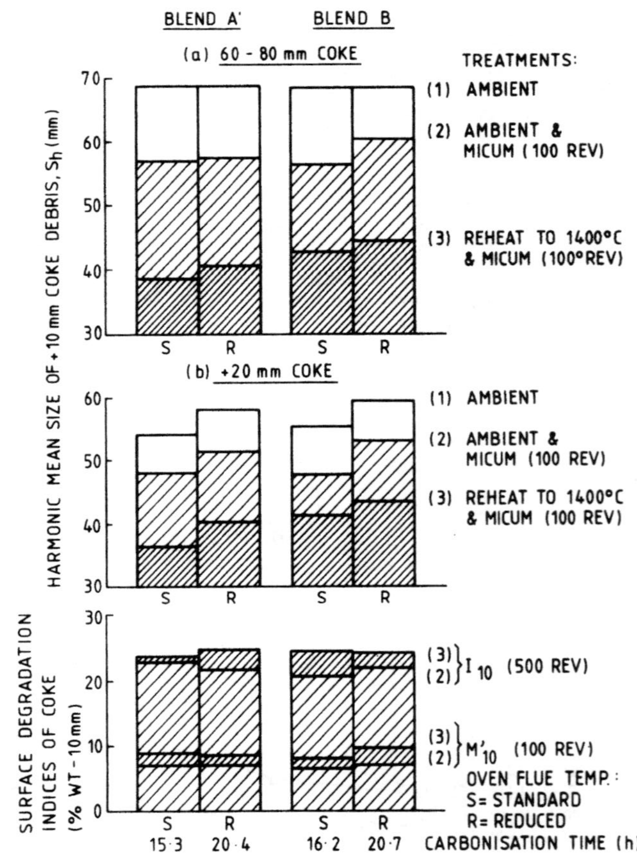

FIGURE 7. EFFECT OF CHARGE CARBONISING RATE THERMO-MECHANICAL STABILITY INDICES OF COKE.

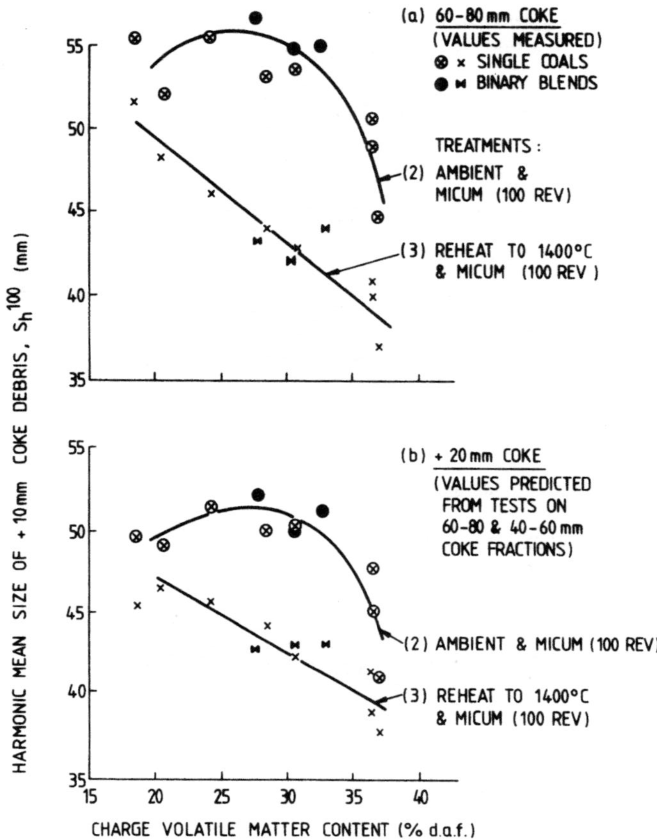

FIGURE 8. EFFECT OF VOLATILE MATTER CONTENT OF OVEN CHARGES ON THERMO-MECHANICAL SIZE STABILITY OF COKE.

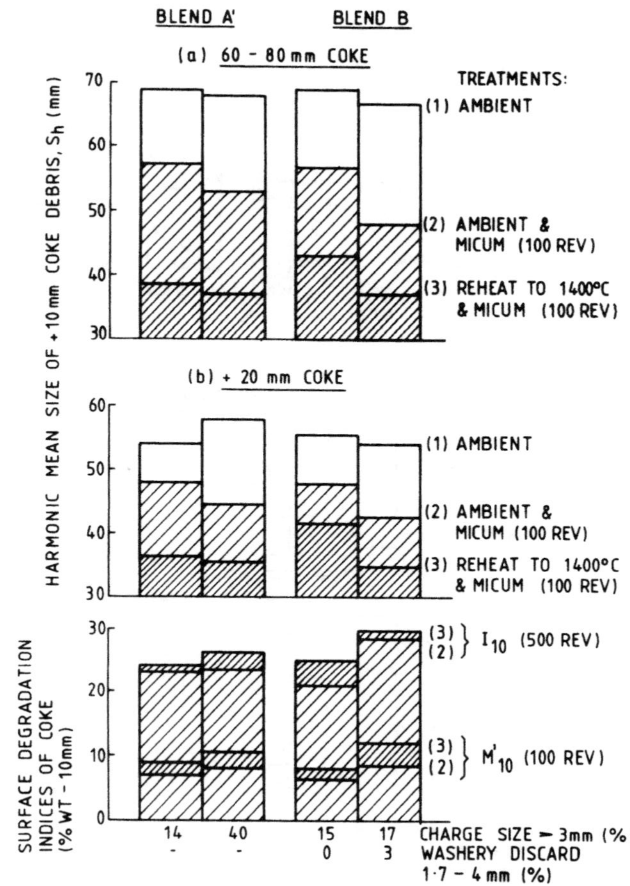

FIGURE 9. EFFECT OF OVEN-CHARGE SIZE CONSIST ON THERMO-MECHANICAL STABILITY INDICES OF COKE.

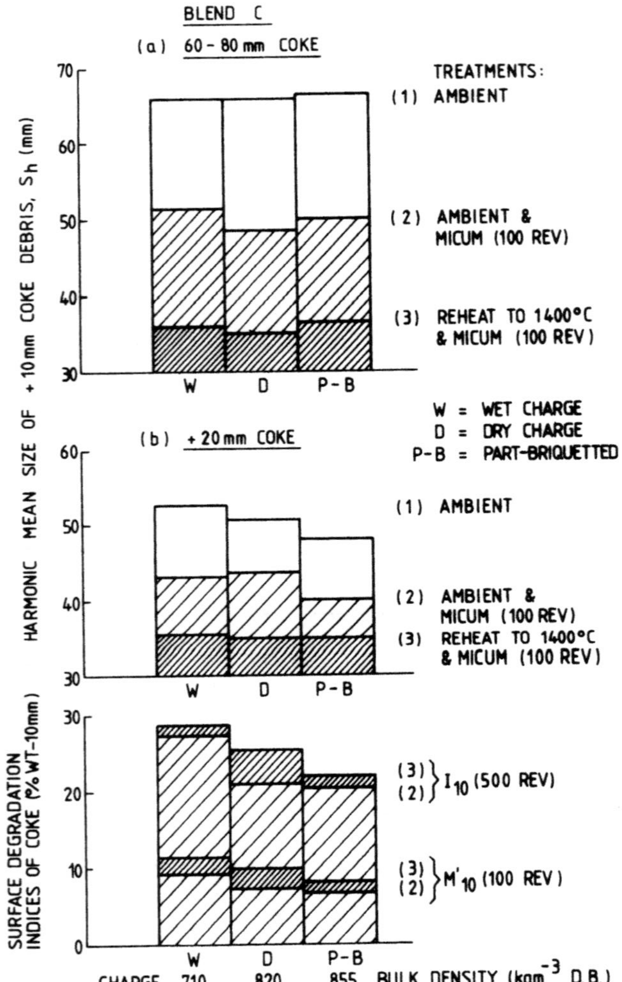

FIGURE 10. EFFECT OF CHARGE DENSIFICATION ON THERMO-MECHANICAL STABILITY INDICES OF COKE

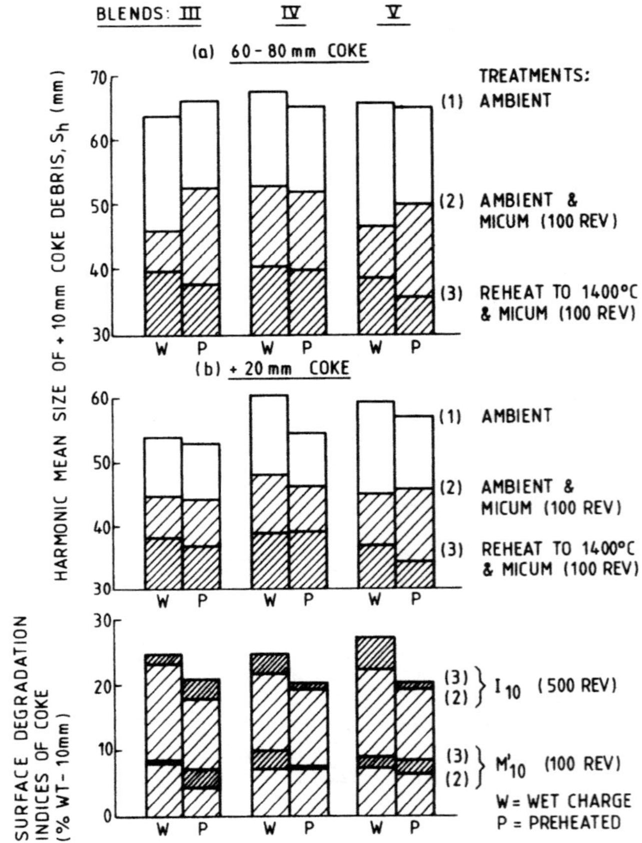

FIGURE 11. EFFECT OF CHARGE PREHEATING ON THERMO-MECHANICAL STABILITY INDICES OF COKE FROM 17t OVEN

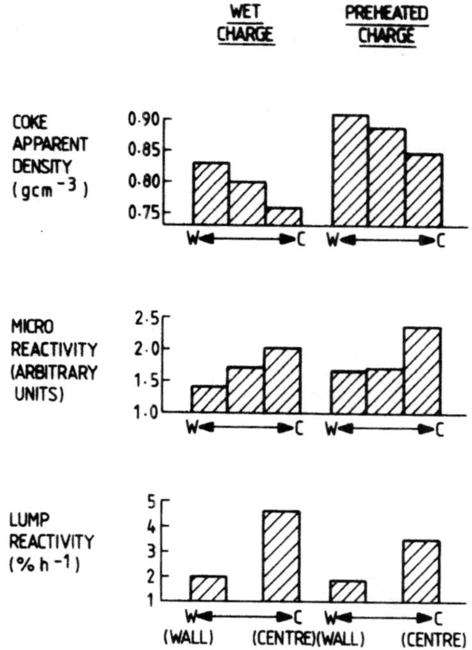

COKE REACTIVITY MEASURED IN CO_2 AT 975°C

MICRO..... 0·1g, 0·125 × 0·212mm

LUMP ..... 5–10 × 35g, 38dia. × 38mm

FIGURE 12. EXAMPLE OF DENSITY AND REACTIVITY VARIATIONS WITHIN COKE MASS (ACROSS HALF-OVEN WIDTH)

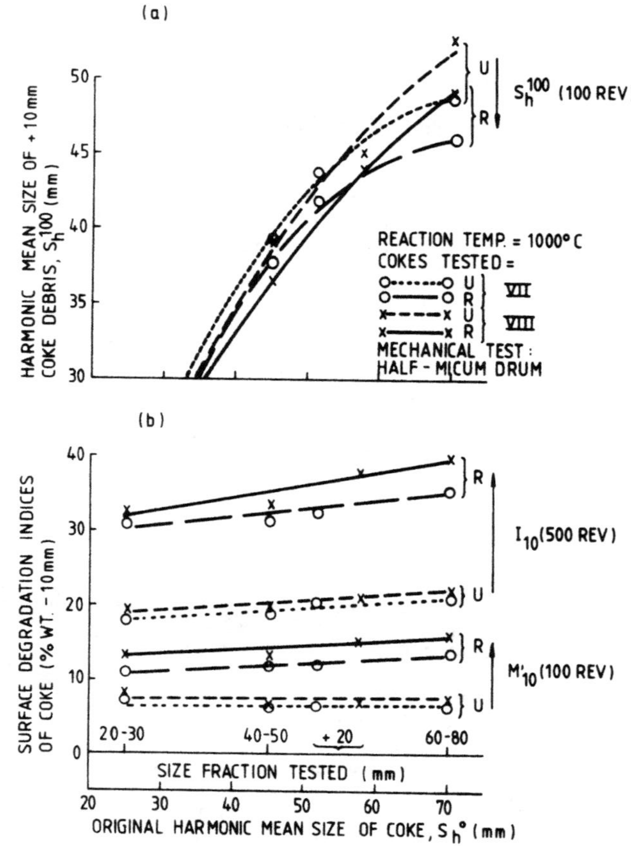

FIGURE 13. EFFECT OF ORIGINAL LUMP SIZE ON DEGRADATION RESISTANCE OF COKES BEFORE (U) AND AFTER REACTION (R) WITH CO_2 TO A CONSTANT WEIGHT LOSS OF 20%.

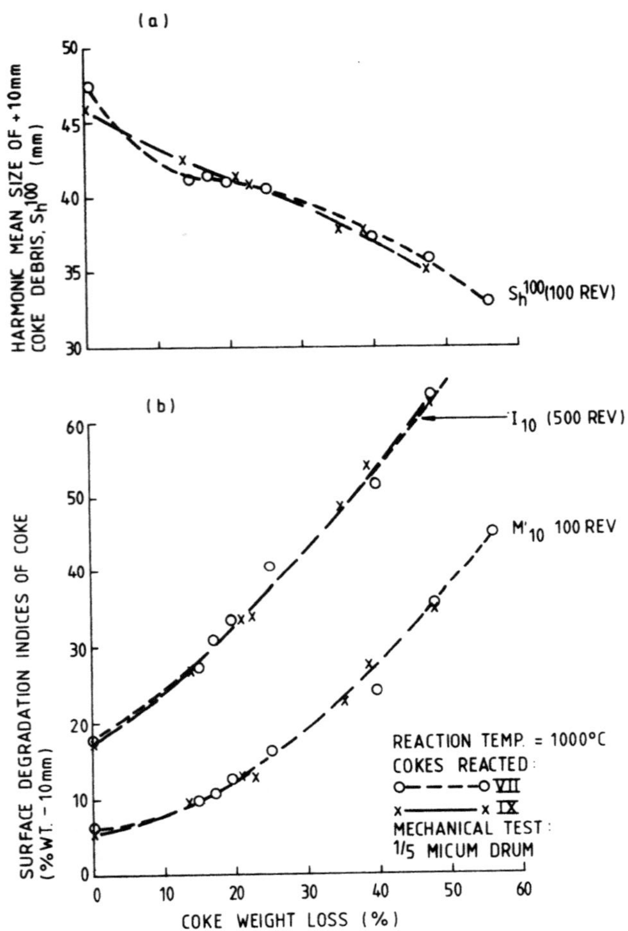

FIGURE 14. EFFECT OF EXTENDED REACTION WITH CO_2 ON DEGRADATION RESISTANCE OF COKES

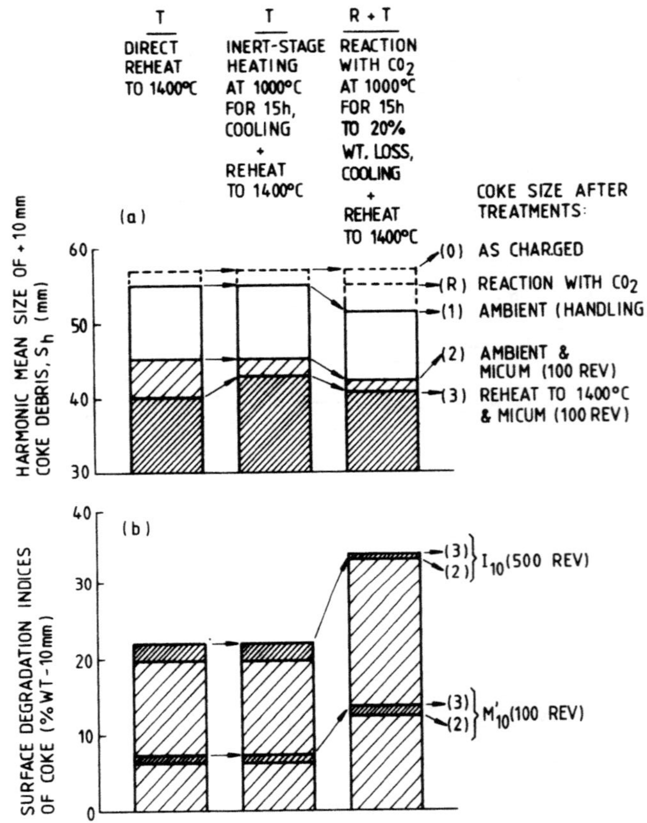

FIGURE 15. EFFECT OF COMBINING THE BULK THERMAL STABILITY TEST (T) WITH PRIOR REACTIVITY TEST (R) ON MECHANICAL STABILITY INDICES OF COKE IX.

CHARACTERIZATION OF COAL AND COKE

By J.M. Duchene and G. Prudhon

Marienau Experimental Station

Summary

The manufacture of coke to meet given requirements presents the problem of the blend formula for the coal charge and more often than not the answer lies in the utilization of experimental correlations between the properties of the constituents and the usual mechanical strength indices of the coke. As regards coal characterization we have to distinguish between identification tests in relation to origin and genesis and evaluation tests which are more directly linked to the use and the utilization capacity of the coal. Petrographic analysis belongs to the former of the two, although it is being used more and more for the purpose of monitoring consignments and for forecasting coke quality, as a complement to the conventional types of analysis aimed at determining plastic properties. In practical terms, the uniformity of the quality of the supplies is an important factor and we know that coking coals are very sensitive to atmospheric oxidation, which may occur during transportation or stocking. We cannot always be one hundred per cent certain that changes in plastic properties and chemical composition are a result of oxidation. We solved this problem by developing a simple, original test, applicable to the whole range of coals and which can be used in the control laboratories of coking plants.

Coke characteristics depend primordially on the nature of the coal being charged and the carbonization conditions. We also know that it is possible, regardless of the blend involved, to have an effect upon conventional mechanical strength indices by altering the coking parameters or changing the method of charging. However, the increasing interest of the blast furnace operators in the behaviour of the coke when hot, has given rise to new and numerous research projects in this field. As our contribution, we developed, in conjunction with the IRSID and with Community aid, a dual-purpose gasification test, aimed at determining the temperature in the reserve heat zone and assessing the coke's mechanical strength on arrival in the blast-nozzle zone. The results obtained are interpreted in the light of the properties of the original coal feedstock and in terms of manufacturing factors.

1. INTRODUCTION

Choice of coking coal supplies is generally based on the technical, commercial and local factors involved. In this context, the coke manufacturers endeavour to produce a type of coke which has a chemical composition and a mechanical strength compatible with the required blast furnace specifications whilst also seeking maximum uniformity over a period of time. In order to meet these objectives the coke manufacturers usually rely on experience and experimental correlations drawn up between the laboratory characteristics of the types of coal they receive and the conventional mechanical properties of the coke.

In this connection it is difficult to forecast and even more difficult to rectify in good time, any unexpected variations in the coking properties of one or more of the constituents of the coal charge. There are two possible reasons for alterations in the new coal consignments: a change in the nature of the coal which occurs in the washery or during shipment or a change on account of atmospheric oxidation. Because of deterioration of the conventional plastic properties it is not possible to resolve which one of these two possibilities is in fact the case. Determination of the coal type and rank by means of petrographic analysis permits control of the origin of the new consignments, but it does not facilitate detection of slight or moderate coal oxidation. The methods of humic acid analysis and analysis of the oxygenated functions are not very sensitive and their usage implies a knowledge of the properties of the fresh, reference coal, which is mere guesswork when the new consignments originate from distant countries. The Marienau Experimental Station has arrived at a practical solution to the problem of detecting coal oxidation, by developing a simple test and the procedure is very close to that of classical coal tests and can be used under current manufacturing conditions.

In another connection, the attention of the blast furnace operators was drawn to some aspects of coke behaviour in the lower part of the blast furnace where temperatures are very high, by the achievements of the Japanese steel industry which was then in full expansion. It rapidly became apparent that the only normal mechanical strength criteria were no longer adequate to determine the behaviour of the coke in that part of the furnace. Numerous research projects were undertaken with a view to determining and testing experiments on the hot coke which best simulated the thermo-chemical attacks experienced by the fuel during its journey downwards through the blast furnace, with special attention being paid to the effects of these changes on the mechanical strength of the coke on arrival in the blast-nozzles zone. The Marienau Experimental Station with the assistance of the IRSID developed a new dual-purpose, hot coke test. It aimed at determining both the temperature of the reserve zone in the blast furnace and the mechanical strength of the coke in the blast-nozzles zone. This test does not take into account the mineral charge, with the result that it is possible to isolate the effect of the coking factors alone. The metallurgical aspect of the problem is dealt with by other methods at the IRSID.

II. DETECTION OF COAL OXIDATION

II.I. Principle and description of method

The fundamental principle of the method is based on the hypothesis that the fixation of the atmospheric oxygen to various types of coal is, generally speaking, a surface phenomenon. Consequently the finest particles are somewhat more oxidized than the larger fractions and graded sizes of coal are covered by a fine film of oxidized coal, the thickness of which is dependent upon the extent of the phenomenon. This fact prompted the principle of the oxidation-detection test which we have developed. Stage I of the test procedure consists of subjecting a fraction of the graded sizes in the consignment of coal to be monitored, for example the coal between 3 -·5 mm, to an abrasive treatment in a small revolving drum; there are holes 3 mm in diameter in the walls of this drum. Stage 2 involves the monitoring, at various stages in the abrasive treatment, of the plastic properties of the fines fraction below 200 microns, collected in the abraded product. The properties of the abraded coal are then measured by means of a Chevenard-Joumier dilatometer which has various advantages compared with the international dilatometer: it takes a smaller test sample and is more sensitive in the range of lower-swelling values. Trends in properties of the intermediate fraction, 0 - 2-3 mm, formed within the drum provide further information on the behaviour of the coal.

If the coal is oxidized, the abrasive treatment removes the film of oxidized coal during the first few rotations of the drum, so that a blend of oxidized coal and fresh coal is then obtained when the non-oxidized parts come to the surface. If the abrasive treatment in the drum is then continued one should in theory be dealing with pure coal again. Consequently, the significant variation in the swelling of the fines fraction during the abrasive treatment enables us to conclude beyond a shadow of doubt, independently of our knowledge of the properties of the fresh, reference coal, whether we are dealing with oxidized or non-oxidized coal. Determining the degree of oxidation is a very difficult task and we have not as yet arrived at a successful conclusion.

In the case of non-oxidized coal, swelling of the fines fraction produced does not vary significantly during the abrasive treatment.

II.2. <u>Interpretation of the results</u>

By applying this test firstly to fresh coal and then to this same coal when it has been subjected to oxidation in a drying chamber for increasingly longer periods of time, curves are obtained similar to those plotted at Figure 1. These curves show that the final swelling variation varies in accordance with the state of the coal: if it is fresh, the swelling remains virtually constant throughout the abrasive treatment, whilst it increases steadily in the case of oxidized coal. The amount of wear to which the coal must be subjected before the increase reaches a perceptible level varies with the intensity of the oxidation.

The swelling trend of the intermediate fraction generally speaking confirms those observations made about the fines fraction and provides further details of the extent of the oxidation, in the case where the latter would have had a mass effect. It should be noted that it is frequently impossible to get back to the swelling figure of fresh coal, even if abrasive treatment is prolonged. This merely signifies that oxidation is not just a surface phenomenon, but that there may be a marginal penetration of oxygen towards the centre of the graded sizes of coal.

II.3. Practical applications

Since the conception of this method, it has been applied to many different types of coal of all origins, whether fresh or oxidized. Figure 2 presents some typical examples.

Experience has shown us that totally fresh coal does not actually exist as such: even in the optimum case, there is a slight increase in the swelling of the fines fraction during the initial abrasive treatment, which confirms the presence of an oxidized surface layer.

Furthermore, this initial film appears to play a protective role, to a certain extent, since generally speaking oxidation in a drying chamber of coal which has already been altered, proves more difficult than oxidation of fresh coal.

On the whole, all homogenous coals react normally to this test, but with a sensitivity particular to each constituent. It is generally possible to raise the swelling back to somewhere close to that of fresh coal. Because of this it is virtually impossible to define the degree of oxidation in quantitative terms.

When dealing with heterogenous constituents, the method can still be applied provided the results are verified by petrographical analysis, as distorsions could occur stemming from preferential abrasion of one series or another.

Despite the various interpretation difficulties which may

crop up from time to time, this test is a very reliable one and one of its main advantages is that one does not need to know the characteristics of the fresh coal. Of course, the result is not qualitative, but as regards operations on an industrial scale, it may make it possible to adapt the formula of the coal charge beforehand or to explain possible anomalies observed during coking or in the mechanical strength indices of the coke afterwards.

III. SIMULATION OF COKE BEHAVIOUR IN THE BLAST FURNACE

1. Principle and description of method

In the introduction we saw that the hot coke test had two objectives:

- to determine the temperature of the reserve heat zone.

This temperature is determined by the onset of the endothermic reaction of carbon monoxide regeneration ($C + CO_2 \quad 2CO$). This plays an important role as regards operation of the blast furnace since it controls the kinetics of iron-oxide reduction and consequently governs the operating condition and efficiency of the two exchangers (upper and lower) in the apparatus.

In the laboratory test, we reproduce blast furnace conditions as best we can by using coke of graded sizes (between 20 and 30 mm), by oxidizing the coke in a heterogenous stream of gas containing nitrogen, carbon dioxide, carbon monoxide and hydrogen and by gradually subjecting the coke to an increase in temperature from 700°C to 1200°C, at a rate of 200°C /hour. The reaction threshold is determined by continuous measurement of the coke temperature and weight simultaneously during the test. The reaction threshold is the temperature corresponding to the appearance of a decrease in the mass of the sample which is measured by means of a thermobalance.

- to control the mechanical strength of the gasified coal up to a weight loss of 25%.

The fact that it incurs a constant weight loss is the second original feature of this hot coke test. One can in fact calculate that, whatever the type of coke used, for a given mineral charge and a blast furnace which is operating correctly, the carbon loss is, at a rough estimate, constant when the coke reaches the zone around the blast-nozzles. These provisions make it possible to compare coke of different origins, regardless of the apparent reactivity of such coke. The use of a thermobalance enables us to define the very moment at which the weight loss is 25%.

On conclusion of the gasification test the coke is cooled to room temperature. A 200 gr sample of the gasified product is then subjected to abrasive treatment in the drum. The abrasive strength is characterized by the proportion of dust formed under 3 mm.

Main results

a) Comparison with blast furnace

On several occasions, the gasification test was carried out on coke samples taken from the charge from blast furnaces operating under normal running conditions. In the case of some of these coke samples, the ordinary monitoring arrangements were complemented by special temperature measurements in the chamber and on each occasion, coke was taken from the blast-nozzle zone. Some examples of the results obtained are given in the following table:

Sample type	Temperature of reaction threshold in °C	Abrasion % <3 mm		
		untreated coke	treated coke	coke from blast-nozzle zone
RUHR	945 955	6.0 6.1	17.0 18.3	2.2 3.0
LORRAINE	935 935 955	6.1 8.3 7.1	22.0 24.2 27.0	3.8 5.3 7.2

Gasification threshold temperatures are always close to 940 - 950°C. In the industrial tests we obtained reserve heat zone temperatures of 950°C in the case of the Ruhr coke and 930°C in the case of the Lorraine coke. This conspicuous degree of concordance shows that the simulation adopted is sound.

Naturally, it is not possible to compare in absolute terms the abrasive strength of the gasified coke with that of the coke sample from the blast-nozzle zone, since the static treatment in the laboratory apparatus cannot reproduce the mechanical stresses experienced by the coke in its journey downwards. However, whether one takes as one's criterion the final production of gasified coke or that of coke from the blast-nozzle zone, the classification obtained is the same, whilst it would be different in the case of the strength of untreated coke.

These few results show that the laboratory test is capable of providing valuable data on the behaviour of the coke in the blast furnace and as a result it must make it possible to study the incidence of carbonization factors on the temperature of the reserve heat zone and the mechanical strength of the coke as it reaches the blast-nozzles.

b) Effect of manufacturing factors on the reaction threshold temperature

We have seen how the types of coke mentioned above all began to react at a temperature close to 940 - 950°C despite the fact that they were prepared from coal of distinctly different types. The same threshold value is encountered in the case of experimental coke prepared in a 400 kg oven whether we alter the composition of the coal charge, the carbonization rate or the method of charging. Only the degree of carbonization (Figure 3) appears to have a significant effect provided the temperature has not been raised above 850°C. The reaction threshold temperature seems to be a constant of the classical coke oven.

We are currently conducting tests with a view to demonstrating those parameters which affect the gasification threshold.

In another connection, the only known types of coke to react at a low temperature are those which are only slightly carbonized and prepared under special conditions.

c) Effect of manufacturing factors on the strength of the gasified coke

It should be borne in mind that the strength of the gasified coke is measured by constant weight loss and that this method eliminates effects which could be due to various other reactivities.

Figure 5 shows that the variation in the strength of coke manufactured from different blends of coal follows different lines depending on the carbon loss of the gasified product. In particular, embrittlement corresponding to a gasification rate of 25% is greater in the case of coal with a higher volatile matter content.

Then again, other coking parameters such as the method of charging, the carbonization rate or the degree of carbonization have virtually the same effect on the strength of the untreated coke as on the strength of the gasified coke.

The mechanical behaviour of the gasified coke depends more, therefore, on the type of coal being charged than on the other manufacturing factors. This remark would suggest that the mechanism of chemical etching depends on the structure of the carbonized phase of the coke which we know is directly related to type and rank of constituents. On a microscopic level, the walls of the pores are gasified at higher or lower rates according to the local reactivity of the carbonized phase. The mechanism of chemical etching, therefore, depends more on the relative size and distribution of different carbonized seats of reactivity than the average or overall apparent rate of the reaction which in this instance has no tangible significance.

The coke's reactivity varies locally with the type and rank of maceral. To a certain degree the oxidation of the coke attacks the carbonized phase of the walls of the pores leaving the impression of the texture of the original coal. In particular, the degree of local oxidation, for a given overall weight loss, will be increasingly smaller as the anisotropy or the degree of organization of the carbon is more pronounced.

IV. CONCLUSION

The coke industry in our country is becoming more and more dependent on supplies of coal from abroad and variations in quality are on the increase and consequently it is at times difficult to know the precise origin. Under these circumstances lies the risk that the coking properties may change on account of atmospheric oxidation during transport and stocking, where the coal may stand for long periods. It is very difficult to define coal oxidation in absolute terms. We have developed a simple test which can be applied along with other current manufacturing controls, which detects whether the coke is oxidized or not. The findings of this test, which can be applied to all grades of coal, may be used in a number of different ways: to determine the requisite adjustments to the blend formula, to optimize the turnover periods in the stockyard, to explain operating anomalies such as poor shrinkage, graphitization and increase in calorific consumption, which could be connected with oxidation.

As trends in blast furnace technology require an increasingly greater knowledge of the behaviour at high temperature of the fuel being charged, the Marienau Experimental Station has developed, with the assistance of the IRSID, a test which simulates the behaviour of coke in an industrial drying chamber. This test could be used in current manufacturing controls, once we have interpreted the information thus supplied into useful practical data for blast furnace operations. We are at present monitoring consignments of coke for industrial-scale blast furnace tests, in order to find correlations between properties of the hot coke and operation of the plant.

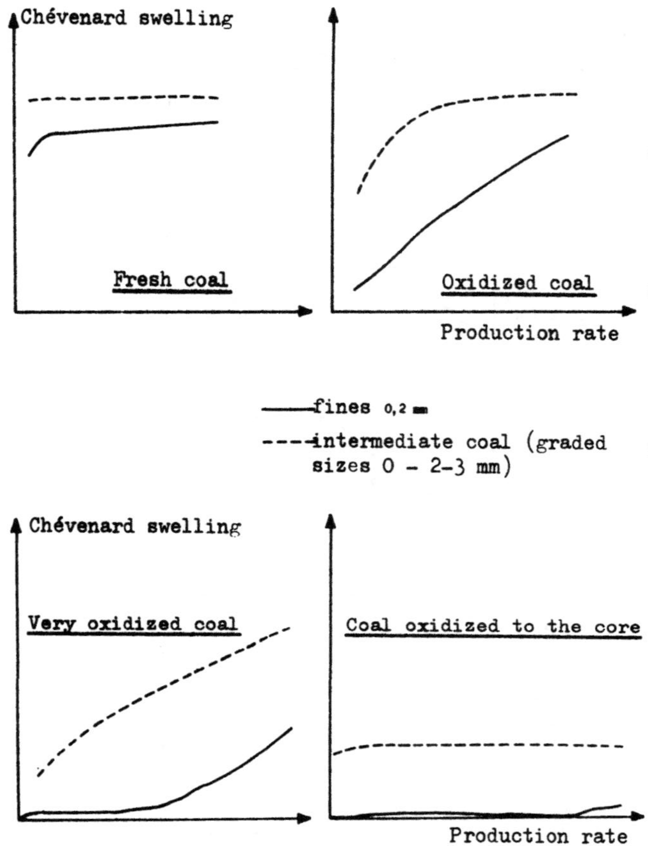

Fig. 1 : Various types of characteristic curves.

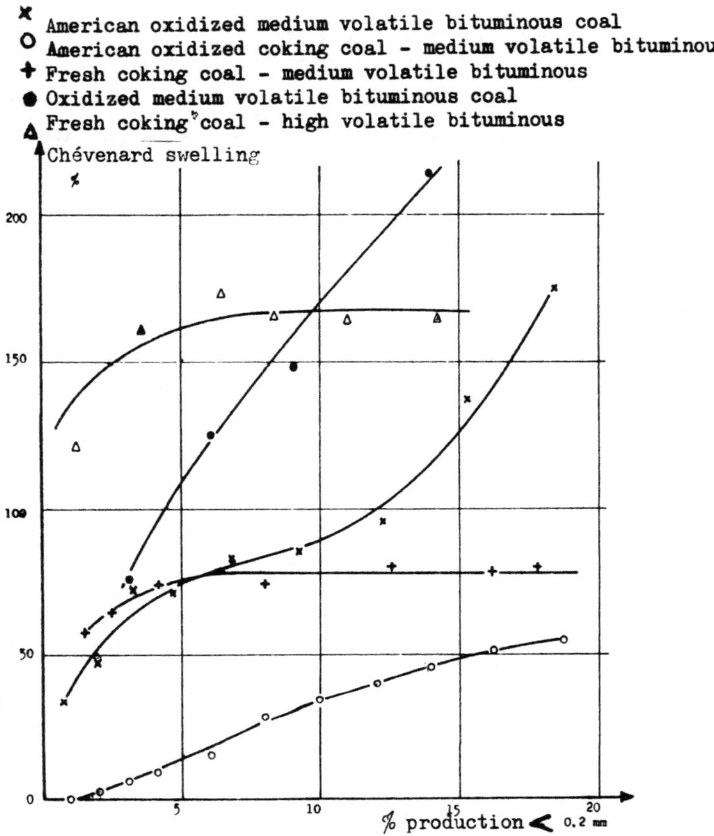

Fig. 2 : Final swelling in terms of production rate for different types of coal

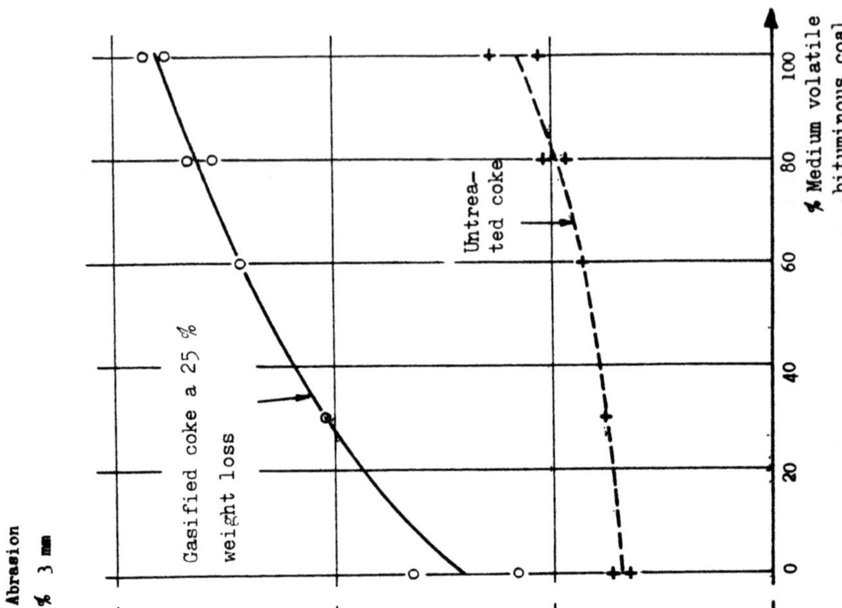

Fig. 6 : Effect of addition of low-grade coal

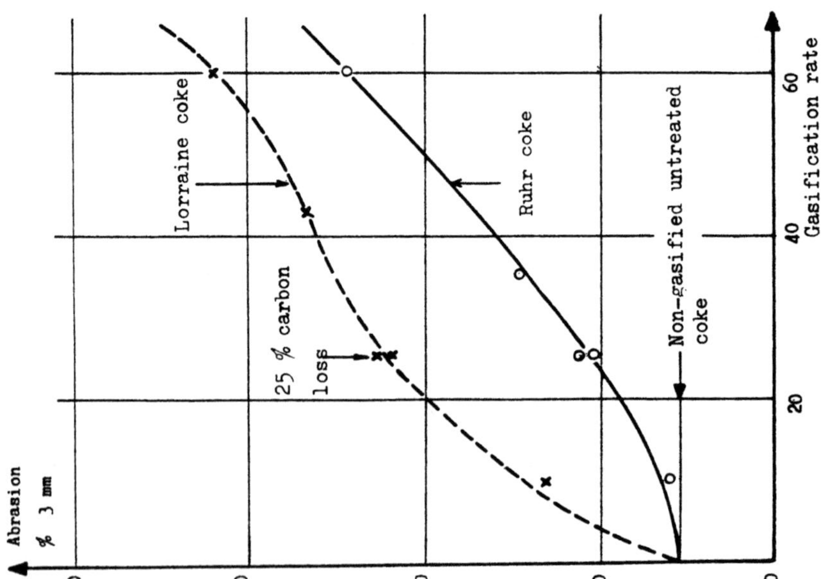

Fig. 5 : Effect of gasification rate

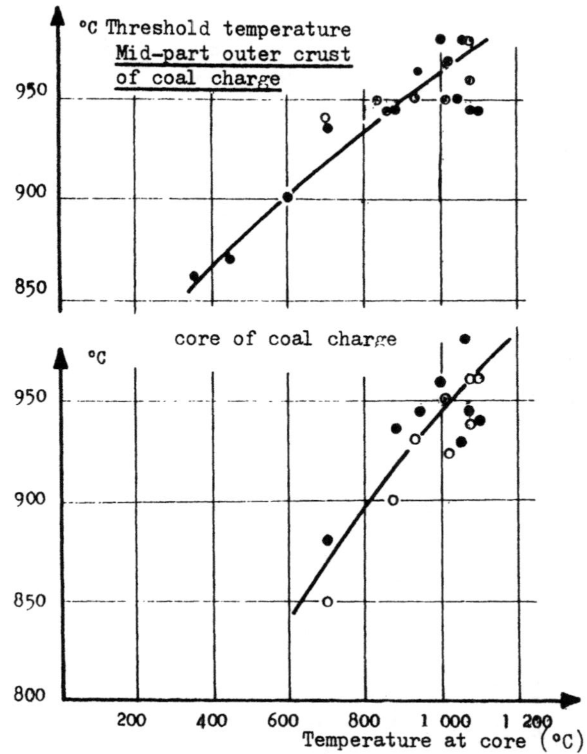

Fig. 3 : Variation of threshold temperature as a function of the degree of carbonization.

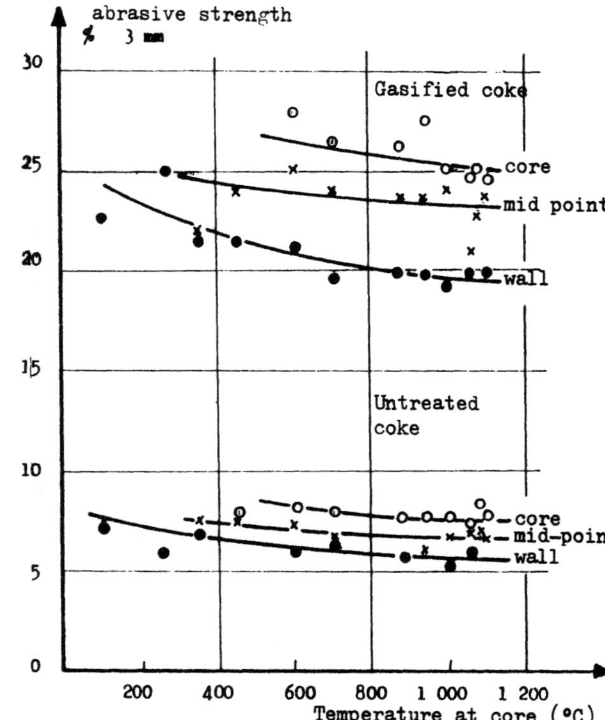

Fig. 4 : Variation in abrasive strength as a function of the degree of carbonization.

ROUND TABLE "A"

A COKE AND THE BLAST FURNACE

Chairman: G. SIRONI

Chairman

Now that all the papers for the first session of this Information Symposium on Coke Oven Techniques have been presented, we shall open the Round Table on "Coke and the blast furnace". However, before we start I should like to make a few remarks which may provide topics for discussion.

The figures quoted in Dr Reichert's opening address show that the current level of coke production in the EEC stands at around 6 olo million tonnes/ year, a large part of which goes to blast furnaces.

In the Community, average specific consumption of coke in this process, discounting the injection of fuel oil, is around 51o - 52o kg/t of pig iron, with substantial differences from country to country. One important point which could be tackled during the discussion is the reduction of the specific consumption of coke in the blast furnace, because of the topicality of the problem both from the point of view of the price rises which followed in the wake in the oil crisis and from the point of view of the performance of the Japanese steel industry.

Another interesting aspect, closely bound up with the papers presented during the first session of this Symposium, concerns the development of coke production technology in the light of the basic studies carried out on the structure of coke: its porosity, and the thickness of its sides. I should also like to stress the relationships between coke characteristics at high temperature and blast furnace operation and I feel that new tests similar to those carried out on ferriferous ought to be carried out to dtermine coke quality.

Another subject of interest is the problem of new coke production techniques, a subject which is dealt with in a specific session of this Symposium. I refer to those processes which use non-coking coals and apply the tech niques of briquetting and formed coke; these techniques are today marking time for a variety of reasons not the least of which is the current plight of the steel industry throughout the world.

Without going beyond the confines of these discussions, consideration could be given to the implications of the characteristics of these new cokes for blast furnace operation.

Lastly, I feel that the debate should not neglect the development of innovations in conventional coking plants, as a basis for a certain degree of process automation.

Let us now open the discussion. Would those interested please note that they may, if they wish so, put questions to the authors of the papers so far presented.

Mr Karboviac, Cerchar (F)

I should like to ask Prof. Abel which particular additive characteristics make it possible to identify those which have an efficient or optimum effect on the plastic properties or the swelling index of a given type of coal.

Mr Abel, Technische Universität Clausthal (D)

In the case of low-volatile (approximately 17-18 %) coals, the swelling index does not play a major role, unlike coals which have a 7o-8o % agglutinating power.

It goes without saying that we have calculated the increase in volume as suggested by Bone Wheeler and Fischer but it has not been possible to bring out differences between the characteristics of the additives.

Mr Marsh, University of Newcastle (GB)

A lot of work has been done and published in "Fuel" on the influence of additives in the coal carbonization process. A large number of additives have been used and analyzed. However, it has not been easy to correlate the behaviour of the additives in the carbonization process with the chemical properties of the additives themselves. It has been noted that hydrogenated derivativs are effective modifiers. One solution to this problem could be to analyze the structural composition of the molecule or, better still, of a part of it.

The aliphatic parts of the molecule may play a significant role. The concept of hydrogen shuttling was recently introduced into the discussion on the development of mesophases during the co-carbonization of coal and pitch.

The modification of the carbonization processes by adding pitch is thus probably linked with the hydrogen shuttle caused by the transfer of hydrogen from the pitch to radicals during the carbonization process. The stabilization of radicals, which, moreover, are apparently reactive, by means of hydrogen capture enables the pitch to interact chemically with the coal to create ultimately nematic liquid cristals which are responsible for the structure required of metallurgical coke.

Mr Beck, BBF (D)

The influence of alkalis on the gasification of coke in the blast furnace is well known. They act as catalysts. Mr Marsh this morning mentioned alkaline aggression affecting blast furnace coke. Was he referring to the increase in the coke's reactivity to CO2 or did he mean direct aggression? Could he elaborate on this or give further details?

Mr Marsh

I should first like to draw a distinction between two aspects. Alkaline metals such as K and Na are efficient calalysts for the gasification of coke They favour the reaction of the CO2 with the C, ie the so-called blast furna ce solution loss. This is the most familiar facet of the problem. Alongside this comes a second point which concerns the behaviour of coke at high temperature in an inert atmosphere, ie free from CO2, H_2O and O_2. In this case the alkaline compounds, eg oxides, hydroxides and carbonates, act as disintegrators of the carbonaceous materials even though there is no reactive gaseous phase. This attack is more often than not observed in respect of the larger dimensions of the anisotropic components of the coke's optical structure. The smaller dimensions and isotropic materials are less reactive. It can be observed that not all alkaline metals behave in the same way; K is more active than Li and Na and the other alkaline-earth elements having a higher molecular weight.

Not so familiar, and still being discussed, is the mechanism whereby K reacts.

But if it is found that the K diffuses from the surface of a piece of coke to the center of the particle, it is feasible that, as a result of this migration, it becomes lodged between two strata and causes the graphite present in the coke to expand and thus disintegrate.

Mr Flasche, Stahlwerke-Röchling-Burbach (D)

I should like to ask Mr Abel whether he has carried out coking tests using blends of low-volatile coal with coal and oil pitch and what correlation he has observed with the results obtained in the laboratory.

Mr Abel

No coking tests or coke strength measurements were carried out. The objective of the work carried out was to clarify those phenomena which could be studied analytically. The underlying aim was to establish a basis for further stages of the research using apparatus such as the Jenkner retort. I should like to ask Mr Marsh whether when using potassium carbonate as a catalyst in the Boudouardreaction he observed that the catalytic effect commenced as from carbonate concentrations of the order of 3 - 4 %. This question relates to similar experiments on gassification using steam and the fact that part of the K is deactivated by the mineral substances in the coke.

What is the extent of variation of gassification recorded in the case of the Boudouardreaction, bearing in mind that in the case of steam gassification increases in the reaction speed of the order of 55 - 65 % have been observed with alkalis?

Mr. Marsh

There are certain discrepancies in the specialist literature as regards the influence on the elements which catalyse the coke gasification reaction, partly because in the past insufficient attention has been paid to the distribution pattern of the catalysts. Sometimes quantities of the order of 10% on the surface of the coke remain inactive, while on the other hand the presence of a few ppm has increased the reaction speed by a factor of 100.

Generally speaking, it is not easy to bring about catalysis in the presence of large lumps, and that is why the percentage weight of additives and the degree of catalytic reaction do not seem to be correlated. From the slightly more scientific point of view, it could be stated that the small particles (e.g. particles of 7 mn) possess fairly different physical properties from those of normal-sized particles, and catalysis makes the actual behaviour different from that expected under the laws of thermodynamics.

When materials are well divided and well distributed in the coke, the Boudouard reaction is greatly affected by alkalis. For a better understanding of the problem, I therefore think that account should be taken both of the characteristics of the coke and of the distribution of the catalysts, Furthermore, if K is added to the material derived from coal (coke), part of it may interact with the other compounds present and thus give rise, for example, to potassium silicate. The result of this is that part of the K no longer acts as a catalyst since it has been eliminated from the system. The mechanism is consequently very complex to examine in detail.

Chairman

Could Mr Marsh state what effect ash has on the gasification of coke? Is there any indication that it acts as catalysts? Is there anyone present able to enlighten us further on this subject?

Mr Marsh

Most of the mineral components present in coke in a fairly high percentage have no catalytic function.

If the reactivity of metallurgical coke is examined and compared with that of pure carbon, no major differences are noted for an equivalent surface area. For instance, the reactivity of metallurgical coke is slightly lower than that of petroleum coke which can be considered a comparatively pure material.

Moreover, metallurgical coke is much less reactive than lignite, or other low-ash coals purpose prepared in the laboratory.

Which is just as well, for if the ash did have a major catalytic effect it would be impossible to get the coke to the bottom of the blast furnace!

Mr Beck

We are focussing our attention on coke reactivity, which rises spectacularly in the presence of alkalis. Blast furnace research such as that carried out by BSC has shown that alkalis from other components in the burden diffuse towards the coke and affect its reactivity. There are only limited possibilities for reducing the reactivity of the coke during manufacture; moreover they are expensive and thus make the coke itself more expensive. What possibilities do the blast furnace specialists see for reducing or eliminating the increase of coke's alkali content in the blast furnace? Would not this be the best solution?

Mr Monson, BSC (GB)

I agree with what Mr Marsh said on the role of alkalis and the effects of ash content. If coke taken from the tuyères is compared with coke as it is at the charging stage, there is no great difference in the ash content. All that can be observed is an increase inthis content due to the effects of gasification. On the contrary, as Mr Beck has pointed out, the alkali content rises markedly by a factor of approximately lo. This entails a substantial increase in coke reactivity between the stock line and the tuyères.

Without dwelling on this subject, my own feeling is that the reactivity of the coke charged into the blast furnace has been overstressed.

The Japanese were the first to note the relationship between this variable and the coke's behaviour in the blast furnace and to measure the reactivity of the charged coke.

This does not mean that I underestimate the importance of the reactivity as regards coke behaviour but an examination of the gasification reactions using CO_2 and H_2O which can occur in the blast furnace shows that more often than not there are no significant correlations between the extent of gasification with reactivity and the quantity of alkalis in the coke charged. It is more essential in my view for the coke to maintain its mechanical resistance to aggression by alkalis rather than for it to react less with CO_2 and H_2O even though it is catalyzed by the alkalis themselves. I refer to the phenomenon of disintegration mentioned by Mr Marsh.

This should be the focus of our attention, rather than the reactivity, which in an event is indicative of certain specific properties of the coke and its behaviour in the blast furnace.

I should like to make one further comment, Mr Chairman, concerning your introductory comment on the reduction of coke consumption in the blast furnace. Praiseworthy as this objective may be, it all depends on the specific configuration of the works involved. Depending on the circumstances, the pursuit of this objective could be counterproductive. For instance, I know of certain works which, because of the demand for steel at that particular time, recently found it advantageous to deliberately increase the coke rate, reducing the temperature of the blast and dispensing with the injection of fuel oil. The purpose was to meet the fuel requirements of the rest of the works by using blast furnace gas.

With the present types of integrated cycle, we are still far from the stage where the blast furnace can be considered a supplier of gas fuel. However, bearing in mind the price of oil, we have to envisage a situation in which the whole works have to function without fuel oil. In this case a coke rate of slightly less than 8oo kg/t of pig iron should still provide enough surplus gas to meet 5o % of the works' electricity generation requirement. I think this figure is too high and that the level of own generation of energy should be limited to levels compatible with essential safety requirements, i.e. probably around 2o % of the total requirement. If the consumption of coke in the blast furnace is reduced beyond these limits use has to be made of alternative fuels and this is uneconomic.

Chairman

I should like to make two remarks. The first concerns the problem of alkalis; clearly, we can not expect everything from coke in this field. A preliminary step, as has been pointed out by others, would be to reduce the specific quantity of alkalis charged into the blast furnace (as I state on page 14 of my report). Furthermore, where build-ups of alkalis occur in the blast furnace it becomes necessary to regulate the basicity of the slag or its chemical composition using techniques which are familiar to blast furnace specialists and which I will not go into now.

As for Mr Monson's remark concerning coke consumption in the blast furnace and the energy aspects of a works, I agree on the need to consider the problem in the steel sector from the economic point of view. Obviously we must not cut energy consumption whatever the cost. Situations may arise in a works with an integrated fuel cycle where saving energy in certain processes means that more costly fuels have to be used in others. However we should not overlook the developments which have taken place in blast furnace technology during the past 2o years.

The notion of not being overly concerned with coke rate in the hope of using the top gas as a fuel could prove dangerous and inadvertently take us back to consumption levels which were a feature of the techniques used many years ago.

Mr Madly, NCB (GB)

There are two facets to the problem of reactivity; one is of a commercial nature and the other of a scientific nature.

BSC specifications for coke stipulate low reactivity levels particularly for large blast furnaces.

Since it is difficult to manufacture coke processing these characteristics, particularly using indigenous coal, the commercial implications are self-evident if, as Mr Monson pointed out, reactivity does not have the importance attributed to it by the Japanese.

From a more scientific point of view, I should like to ask Mr Marsh whether he has noted any reduction in the activation energy of the gasification reaction as a result of adding catalysts or any increase in the number of pores due to the segregation of the coke's structure.

Mr Marsh

I should like to make a general remark and recall firstly that the word "reactivity" is used with the original meaning of speed of reaction as expressed using a single parameter. I was surprised to hear a producer and user of coke saying that his coke was highly reactive and yet perfectly fulfilled its function, and then to hear comments suggesting that coke possessing a low level of reactivity should be manufactured.

So far reference has been made to the single parameter I mentioned earlier. I feel that the way in which coke reacts has not yet been sufficiently studied. In actual fact, a level of gasification of the order of 1 % of the coke can entail drastic effects as regards the weakening of its structure. On the other hand, high-speed gasification may leave the structure intact and reduce only the size of the lump. This accounts for the different opinions expressed by producers or users of coke as regards the reactivity of different types of coke. As for Mr Madly's question, let me say that from the scientific point of view the mechanism of the phenomena in question is very complex, but, if I may, I can simplify this as follows.

In the presence of alkalis, there is a combination between the aggressive action which breaks the structure of the graphite contained in the coke and that due to gasification. If the pattern of activation energy is observed, it will be noted that it diminishes, but that by compensation the apparent number or pores decreases.

It is not therefore enough to possess the kinetic data, which cannot illustrate clearly the reduction in activation energyyand the increase in the number of pores. It is also necessary to look at the topography of gasification in order to demonstrate that the presence of K is accompanied by chemical reaction. In this connection, it can be observed that there is a discrepancy between the kinetic data and the data observed using a scanning microscope.

Mr Monson

In connection with what Mr Madly said, I should like to state that the level of the coke's reactivity at the tuyère is not important. BSC specifications for reactivity refer to coke at the charging stage.

The reason why low levels of reactivity were required is that these give good results in the blast furnace.

Mr HOSTI, Usinor (F)

Prof. H. Guerin, in a bibliography on coke reactivity thirty years ago, listed approximately 5oo references. The catalytic action of alkalis on the Boudouard reaction was well-known (before the Japanese).

As the influence of coke reactivity on the blast furnace is widely recognized, I should like to ask Mr Marsh whether he does not feel it would be advisable for coke producers to exploit this property of the alkalis to improve the quality of the coke manufactured.

Mr Marsh

I think the Japanese deserve credit for having solved the problem of using coke in large blast furnaces and also for having introduced new elements into the European view, particularly the changes in mechanical resistance of the coke after reaction.

Coming back to the question asked, it can be pointed out that Guerin's analysis was followed by other studies on catalysis. An excellant bibliography has already been published by the General Electric Company in America. Opinions do not seem to vary greatly as regards the mechanism where-by potassium salts or other alkaline salts cause catalysis. However, it should be pointed out that coke gasification as such in the blast furnace is not greatly affected by catalysis. I repeat that checking the quantity of alkalis introduced into the blast furnace and of the structure of the coke in order to minimize the catalytic action are the aspects on which attention should be focused.

Some indications were obtained when it was noted that the larger dimensions of the coke's optical structure are more sensitive to aggression by alkalis.

If the question were to be slightly reworded, perhaps the structure of the coke and the mixes could be analyzed with aview to minimizing alkaline aggression.

Mr Vidal, CRM (B)

In reply to Mr Hosti's question on the addition of alkalis to coke in order to control its reactivity, I should like to state that blast furnace specialists explicitly and unanimous ly consider that alkaline oxides are harmful both for blast furnace operation and refractory life and that it is therefore necessary to keep the quantities of alkaline oxides charged into the blast furnace as low as possible.

I should like to take this opportunity of giving my opinion on the influence of coke reactivity on blast furnace operation.

The whole debate on this issue is something akin to the Loch Ness monster. Twenty or so years ago a controversy arose in Germany. During that period coke reactivity occurred in a high temperature range (between 1 000 and 1 300° C). One steelmaking company which was charging iron ore with a low degree of reducibility had noted that as reactivity increased so did the coke rate. Another company operating nearby and using a burden which was markedly more reducible, on the contrary observed that coke reactivity had no effect on the coke rate. In actual fact, the results and the observations were not contradictory; the thermal and chemical configuration of the blast furnace depends on the speed with which iron oxides are reduced and the speed with which the reducing CO is regenerated. The operational patterns of the modern blast furnace mean that the temperature at the beginning of the gasification mor important, i.e. the temperature which determines the temperature level in the blast furnace reserve area.

A fall in this temeparture will lead to reduction in the isothermal zone of the blast furnace and, consequently, in the coke rate, since the degree of oxidation of the gas in equilibrium with the FeO will be higher. In practice however, this saving in coke can be achieved only if the speed with which the iron ores are reduced at low temperature is increased, something which cannot be achieved without modifying other parameters.

As for coke reactivity at high temperature (1 100 - 1 300°C) calculations using a mathematical model show that it no longer has any influence on coke rate in the modern configuration of blast furnace operation.

These considerations concern only coke reactivity and not the phenomena linked with the weakening of the coke structure during gasification, which is another matter.

Mr Eisenhut, St BV (D)

I should like to ask Mr Patrick whether he can explain how porosity, the size of the pores and the thickness of the surrounding coke depend on the temperature of the oven walls and on the pattern of distillation.

Mr Patrick, BCRA (GB)

We have not carried out a direct examination of the influence of temperature variations during carbonization but we do possess some information drawn from small-scale tests.

Generally speaking, they are only slight variations in pore size when solidification temperature is exceeded. Pore structure undergoes very little change up to final temperature. Nevertheless, more information is necessary in order to be check the validity of this statement.

Mr Simpson, BSC (GB)

I should like to shift the discussion slightly to the subject of coke size.

Coke rate per blast furnace is capitally important if the specifications concerning maximum and minimum size are to be respected. Now that the demand for coke fines for sintering is declining, I should like opinions as to the optimum size for blast furnace coke.

In Japan, the size specified is 25 - 75 mm, while in Germany the trend is towards an even bigger size.

Is there a critical fraction for the blast furnace in the 25 - 4o mm range?

Chairman

While we wait for a reply to the question I should like to say that the trend in Italy is towards minimum coke sizes of 25 mm for the smaller blast furnaces and 4o mm for the bigger ones, i.e. towards a certain redaction of the particle size spectrum. The new system of bell-less charging could result in more efficient utilization of particle size distribution. Nevertheless, I do not think that this will enable particle sizes of under 25 mm to be used in the blast furnace.

Mr Abel

The size distribution of coke should not be examined in isolation but in relation to the particle size distribution of the sinter, the pellets and the iron ore lumps. Allowance must be made for some amount of penetration of the ferriferous materials into the coke bed. For instance, taking a large lump with void ratio of 55 %, a minerals penetration of 1o to 15 % of the volume reduces the void ratio to 4o %. With small-sized coke and ore lumps a void ratio of 25 % is possible. It should be noted that the degree of void space enters Ergun's equation to the power of three.

he type of ferriferous material used is also significant. If pellets are used, the tendency is for the void space to decrease, resulting in more material ib the blast furnace but higher charge losses.

Mr NOEL, Iniex (B)

Referring to Mr Duchene's report this morning, I should like to ask whether he was able to establish any relationship between the aptitude of a given type of coal to oxidize and its structure and / or petrographic composition on the one hand and, on the other, its structure and/ or mineralogical composition.

Mr Duchene, Cerchar (F)

The test developed will make it possible to establish qualitatively whether a coal is oxidized or not.

As for the maceral composition we established that the abrasion test does not modify that of the initial coals. Furthermore, with a heterogenous coal or coal blend we established that the variations in the swelling index on the dilatometer are not due simply to a variation of the initial composition.

Mr Abel

As for the studies on the first stages of coal oxidation, i.e. the first 5 mm of the surface, dilatometric methods are not the best suited in my opinien, the plastometer is more sensitive. There is also another method worked out by Polish colleagues, based on dispersion by flotation. An oxidized coal can no longer be floated and remains at the bottom.

I should like to ask Mr Duchene whether he has tried this latter method and whether he has had accasion to compare the results of the abrasion test devloped at the Cerchar and behaviour during flotation.

Mr Duchene

During the work carried out on the determination of oxidation we endeavoured to use a test currently applied in coking plant laboratories, as mentioned in the report I submitted. We have not up to now had the opportunity to make comparisons with other oxidation tests.

Chairman

I should like to bring the rapporteurs who have presented works on the high-temperature behaviour of coke into this. Our discussion so far has been centered on reactivity.

As Mr Vidal stated, reactivity does not always lead to an increase in the specific consumption of coke in the blast furnace. Moreover, in normal operational conditions it is not easy to demonstrate the effects of this characteristic of the coke, even with sophisticated data-processing techniques.

Results submitted at the Middlesborough meeting in 1977 on the theme "Coal, Coke and the Blast Furnace" and experiments carried out by the Japanese show a marked drop in the particle size of coke as it moves down inside the blast furnace.

I should like to ask whether anyone has had the opportunity to demonstrate and describe the reduction of particle size as a result of gasification with CO_2 and of thermal stresses to which the individual coke lumps may be subjected.

Mr Goleczka, NCB (GB)

When I presented my report this morning, I endeavoured to single out the chemical and thermal causes of coke degradation. I stated that the layer of coke at the tuyères is partly due to mechanical degradation as it moves down the blast furnace.

Coke reactivity, particularly its resistance after reaction, is not only the consequence of its initial condition in chemical terms, but also the result of aggression by alkalis which modify the level of reactivity.

A comparision of coke samples taken at the furnace throat and at the tuyères in a number of blast furnaces shows that a reduction of approximately 5o % occurs in particle size. For instance, a 6o mm size at the throat becomes 3o mm at the tuyères. It is difficult to accept that this reduction is due exclusively to solution loss.

I should like to stress that the state of the coke at the tuyères is the result of three parallel actions, viz. mechanical degradation, chemical degradation and, lastly, a thermal degradation of size.

Chairman

We have reached the end of the time available to us and we must therefore bring this lively and interesting round table discussion to a close. Other points could have been discussed in connection with our main theme. I hope they will be taken as the basis for further discussion in an equally interesting forum in the not too distant future.

Tomorrow, when bringing this Information Symposium on Coke Oven Techniques to a close, I will have the opportunity of drawing some conclusions from our discussions.

I thank all for attending, particularly those of you who have taken an active part in the proceedings.

ROUND TABLE "B"

POLLUTION PROBLEMS

chairman: J.P. GRAHAM

SUMMARY OF THE DISCUSSION

Mr. Graham first thanked the Commission for inviting him to undertake the Chairmanship of the Round Table session on Pollution Problems. He suggested that it would be useful to begin by putting the Round Table into perspective in relation to the remainder of the symposium. Only one technical paper, out of a total of fifteen which were to be presented during the two days, was directly concerned with pollution. Remembering some of the comments in the most interesting and balanced paper by Dr. Reichart in the opening session, the Chairman suggested that a somewhat different emphasis might have been expected. It was also relevant to comment that the cost of the provision of the requisite anti-pollution measures on a new coke oven facility was probably of the order of 20% of the total capital costs of the plants, a factor of some concern under present economic conditions in the industry.

At the invitation of the Chairman, Mr. R. F. Littlejohn presented his paper on the "Emission of Tar Fog from Coke-Oven Doors" which he had prepared in conjunction with his colleague at the National Coal Board, Mr. M. A. Wright. In introducing his paper, which is reproduced in full elsewhere, Mr. Littlejohn described how the work had started during 1974, with a preliminary report being presented to a Round Table Discussion held in Luxembourg on 7-8 June 1977.

With the aid of slides, he summarized the salient points of the investigation, suggesting that at one plant up to 99% of the leakage emission arose from the doors and emphasising the reproducibility of visual subjective assessments of smoke leakage. In this context, he mentioned the difference between the uniform method of the British Carbonization Research Association (BCRA) and their technique.

Finally, having described the essentials of the collection and sampling system used to quantify emissions from a single door, Mr. Littlejohn explained how it had been possible, to quantify total door emissions for each of the three plants so far visited. This was accomplished by multiplying the average grade of leakage, assessed visually, by a factor for each of the plants. The results so far obtained, for two plants with self sealing doors and one with luted doors, appeared to show that emissions of visually similar grades of smoke were in fact less for luted doors, although he advocated caution on the acceptance of this particular interpretation, revealing that further measurements were presently being undertaken at another plant with luted doors in order to attempt to confirm the original data.

Thanking Mr. Littlejohn for his presentation, the Chairman invited questions relevant to his paper.

Noting the fact that a figure for tar loss due to door leakage, had been given for one plant, Mr. M. M. Bowness (British Steel Corporation) wondered if a figure had been established for the total by-product loss via leaking doors and the ensuing costs of such losses.

Mr. Littlejohn commented that whilst he believed such a study was feasible,none had been carried out.

Dr. W. Eisenhut (Steinkohlenbergbauverein) gave a brief account of the work which his organization were carrying out at various coking plants in West Germany. Explaining how the on-going investigation into causes of door leakage was supported by the ECSC he mentioned the useful co-operation which had taken place at the outset with workers in the UK. Dr. Eisenhut expressed the opinion that whilst it was not possible to ever achieve a complete seal of oven doors with metal to metal surface contact, it had been shown that the replacement of fixed edge with flexible metal seals reduced emissions by around 65%. Citing the fact that a notch of less than 0.1 mm between sealing surfaces, due to cleaner damage, would seal after about 20 minutes, he stressed the importance of good maintenance and effective door and jamb cleaning. Looking to the future Dr. Eisenhut said that research in West Germany was presently directed towards the use of new sealing materials and investigating the potential of doors with double seals.

Taking up the subject of automatic cleaners, Mr. Bowness mentioned the development of a simple, but very effective, high pressure water jet door cleaner which had been developed at the Teesside works, and later also installed at the Redcar works, of the British Steel Corporation. Being much easier to keep in operation than the mechanical cleaners previously used, it produced an absolutely clean surface on the metal seal, something which was not necessarily required on the jamb surface, where Mr. Bowness believed that a residual film of tar was most advantageous.

Mr. B. P. Kirk (BCRA), commenting on some of the issues raised by Mr. Littlejohn's paper complimented the authors for what he considered was a very useful piece of work. He believed that the most important outcome was the established correlation between the results of visual assessment surveys of smoke leakage and actual Benzene Soluble Material (BSM) emissions. He emphasised the fact that the BCRA monitoring method, mentioned by Mr. Littlejohn, and used uniformly in the UK, had quite categorically not been aimed at quantifying smoke emissions but had been devised with the specific purpose of reflecting standards of operation and maintenance at a coking plant in so far as they affect smoke leakage from certain defined sources. Indeed since the threshold limit value (TLV) on (BSM) of 0.2 mg/m^3 in working atmospheres was the crux of the present environmental pressure in the UK, Mr. Kirk believed that with periodic measurements of BSM in working atmospheres, and regular smoke leakage surveys, the industry had a very useful system of assessing changes in standards of pollution control. At the same time it would enable plant maintenance to be correctly planned and could be used to judge the effectiveness of modified operating practices. In conclusion he mentioned proposals in the UK for guidance notes for the coking industry, issued by the Health and Safety Executive, which were intended eventually to meet the TLV for BSM but set target standards of leakage emissions defined by the BCRA assessment technique.

Dr K. G. Beck (Bergbau-Forschung) considered that, bearing in mind the information given by Dr Reichert earlier on the age of the coking plants in the European Community, it was not surprising that door sealing remained a residual problem. Nevertheless, it was important, he suggested, to remember the problems, such as dust emissions on charging and pushing, which the industry had faced some ten years ago. These problems, like the dust emissions from coal preparation and coke handling, were now practically solved. In order to accomplish this success, Dr Beck believed that the industry had made great efforts and could be pleased with this notable contribution to the improvement of environmental conditions.

At this point in the Round Table, and at the invitation of the Chairman, Mr V. Kiriline (USSR) showed a short film depicting the application and process of dry coke quenching.

Following the film, the Chairman suggested that dry coke cooling was essentially an energy saving process, although it did control the residual pollution which would be associated with the conventional quenching process.

Mr J. Fiedler (Landesoberbergamt N.W.) did not concur with this emphasis, highlighting the quenching tower as a source of sulphur and sulphate emissions and citing the Japanese expectation that all their plants would be operating dry coke cooling within three or four years. Unlike Dr Beck, he believed that many coking plants in West Germany had much to accomplish in the field of pollution mitigation where, for example, he was concerned with the levels of benzene in the general atmosphere surrounding coking plants.

In accepting that problems remained, the Chairman noted that in the UK there had been a definite change of emphasis in recent years away from the problems of dust nuisance to those of the working conditions of plant operators.

Mr L. Hosti (USINOR) summarized a paper which he had recently prepared concerning environmental protection at a thirty year old coking plant in France with which he had been very closely involved. He began by classifying three generations of coking plants. Firstly, those built immediately after the end of the war, secondly, those plants about ten years' old and thirdly, plants built during the last three or four years. He believed that the latter two groups either have built-in solutions to environmental problems or had means available for application which would considerably mitigate the difficulties.

The old plants in the first group presented a considerable problem. As an example of this category, Mr Hosti summarized the various modifications which had been made at one such plant. Improvements had been brought about by modifying the charging procedure to include the use of aspiration steam in the ascension pipes and the provision of aids to coal flow from the hoppers. Cleaning operations had been changed, charge hole lids modified and water sealed ascension pipe caps installed. Explaining how door sealing remained a great problem, Mr Hosti believed that it could only be solved by improving maintenance. Similarly, he felt that the key to the control of emissions during oven discharge was good carbonizing conditions. Concluding, he considered that an 80-per-cent reduction in emissions had been accomplished and believed that whilst the BSM standard was difficult to attain, it was not always exceeded. It was his view that extraordinary improvements had been brought about by a number of relatively minor modifications.

In thanking Mr Hosti for his comments, the Chairman recalled that the environmental authorities in the UK had always placed great emphasis on what they referred to as "good housekeeping" at coking plants as distinct from the more costly capital equipment frequently advocated.

Following on Dr Beck's remarks, Dr G. Nashan (Ruhrkohle AG) also referred to the relative old age of much of the European coking industry, advocating the future installation of a smaller number of larger coke ovens which would effectively reduce the number of potential sources of pollution. He believed that attention should be focused on avoiding the causes of pollution, a subject which he felt should receive more support from the Commission in the form of contract aid to research projects. Turning to dry coke cooling, which he believed was enormously expensive for a slight reduction in pollution, Dr Nashan nevertheless felt that it should be examined as part of a planned energy policy at a coking plant, perhaps in conjunction with coal preheating.

Referring to the TLV's on both BSM and benzene, the latter having been halved yet again in new draft regulations in West Germany, Dr Eisenhut expressed the view that both the public and the environmental authorities alike should be made aware that these standards could never be satisfactorily achieved.

Mr Kirk felt that, whilst a more pragmatic attitude thankfully existed in the UK, the epidemiological evidence obtained in the USA (upon which the BSM standard had been based by retrospective statistical calculations) deserved further justification by the regulatory authorities in the light of the more recent UK epidemiological data which appeared to draw less gloomy conclusions on the whole subject. It was his opinion also that since the priority concern of the coking industry was with occupational hygiene aspects, there was doubtful logic in applying considerable sums of money to the installation and operation of complicated and massive equipment for the control of oven discharge emissions. A study at the BCRA had shown that these emissions contribute a negligible proportion to battery top levels of BSM and could by good operation be minimized to large particles of nuisance dust which would not, at the great majority of plants, present much of a problem. He believed that at the present time most of the money available for pollution control should be applied to the priority problem of improving working atmospheres to a satisfactory level.

In summing up the discussion the Chairman referred to the range of opinions which had been expressed in regard to capital expenditure, the need for atmospheric pollution control standards and indeed the level of those standards. He suggested that it was perhaps very easy to say that the industry was being asked to spend money which it just did not have, on environmental control. On the other hand, he reflected that an increasing number of practicable means of control were now becoming available to reinforce the substantial progress which had already been made. He also detected a shift in emphasis away from areas which could be termed problems of nuisance to those relating to health and dealing with defined atmospheric standards for the areas in which plant operators worked.

After confirming that none of the delegates dissented from these general remarks the Chairman thanked the interpreters and all those who had taken part in the discussion and reiterated his earlier thanks to Mr. Littlejohn for the presentation of his paper.

BPK/WJT
30.9.81

EMISSION OF TAR FOG FROM COKE-OVEN DOORS

R.F. LITTLEJOHN and M.A. WRIGHT

National Coal Board, Coal Research Establishment,
Stoke Orchard

Summary

Measurements at self-sealing and luted doors are compared and the possibility of obtaining estimates of emission via the British Carbonisation Research Association's "Agreed Uniform Technique" is assessed. It is shown that (a) correlation between smoke grade and emission rate is good for individual works but is not the same at each (b) some modification of the B.Ca.R.A. technique is necessary in order to avoid over estimation of emission.

1. INTRODUCTION

Although leakage of tar fog from coke ovens has long been recognised as contributing to atmospheric pollution and as a health hazard to plant operators, quantitative determinations of the rate of emission are relatively recent. Since 1974 CRE has developed and proved sampling equipment and techniques to measure leakage from oven doors, ascension pipes and charge holes. The basis of the method, which is the same for all three types of source, is to grade each individual source visually according to the appearance of the smoke plume resulting from the leak and then to measure emission from sufficient sources in each grade to provide a reliable value of its mean emission rate. Total emission rate is then obtained from the sum of the products of mean emission rates and numbers of sources for every grade. This approach has two practical advantages; all visible leaks are included whether or not they are accessible for sampling and, once mean emission rates have been determined, emission can be monitored by visual grading only.

The initial stages of this investigation were carried out under an ECSC Research Agreement (Ref. No. 6220-EB/8/804) and reported at the Round Table Conference on Coking Plants, Luxembourg, 7-8 June 1977. These observations showed that, for one works, the amounts of smoke emanating from oven doors were much greater than those from charge hole lids or ascension pipes. Measurements showing that the doors contributed over 99% of the total tar emitted confirmed the visual assessment. Subsequently NSF Ltd has supported an extension of the study to two further works. Visual assessments at both confirmed the relative unimportance of leakage from charge hole lids and ascension pipes so effort was concentrated on measuring the rates of emission from doors only.

2. VISUAL GRADING OF SMOKE

The CRE technique for visual grading closely resembles that of the "Agreed Uniform Technique" described in detail in B.Ca.R.A. Special Publication No. 19. However, the two methods, which were developed independently and almost simultaneously, have different objectives and, consequently, some difference in the definitions of grades. The Agreed Uniform Technique is intended to indicate the urgency of repair on the basis of visual grading alone. It treats oven and levelling doors as separate sources and takes account of the number of leaks from each, ignition of gases, etc. in addition to the quantity of smoke. The CRE scale was designed to be a basis for quantitative estimation of emission and therefore considers only the quantity and quality of smoke irrespective of the number of leaks or the parts of the door from which they come. Therefore it is the more easily quantified since, at most works, only ram side doors are accessible and it would be difficult to measure separately the contributions of leveller doors in the presence of large amounts of smoke from leaks round the main door.

Visual grading is essentially subjective. The number of variables that can affect the appearance of smoke is too great and their interactions too complex for precise specification of grade boundaries to be possible. However, this does not necessarily make it unreliable. There are many fields in which equally complex assessments are made regularly by nominally unskilled operatives. For example grading some types of vegetable for market takes account, however subconsciously, of size, shape, texture, density, colour, odour, etc. The criterion for judging such a process is not 'can it be defined?' but 'does it work?'.

Eight surveys on different occasions at one works gave an average grade per door ranging from 1.6 to 1.9 on a scale of 0 to 4. At another works four surveys, also on different occasions, gave the degree of reproducibility shown in figure 1. Another characteristic shown by this figure is that there are relatively few doors in the higher leakage grades. This is to be expected from the changes that occur during carbonisation. Firstly the pressures inside ovens, i.e. the pressure drops across leaks, fall as shown in figure 2, secondly the composition of the volatile matter changes, tars being released mainly in the early stages, and thirdly leaks can become blocked by condensed tar. The combined effect of these factors can be seen in figure 3 which represents all the observations in one survey plotted against time after charging. The incidence of grade 4 leakage is at a maximum immediately after charging, grades 3 and 2 peak after about three and seven hours respectively, while grades 1 and 0 increase in number throughout the carbonising period. It is also worth noting that grade 0 includes a small number of recently charged ovens thus showing that leak free operation is sometimes possible.

3. MEASUREMENT OF EMISSION

The basic requirements of any practical method for measuring emission from coke oven doors are to:-

(i) operate without impeding normal work on the ovens
(ii) operate without changing the rate of leakage
(iii) collect the emitted material without loss
(iv) extract representative samples from the bulk
(v) measure sample and bulk flow rates
(vi) recover all the tars in the sample

The apparatus shown in figure 4 was developed to meet those requirements as far as possible. It is mobile and can be wheeled along the bench on the ram side of the ovens. It can be set up at any chosen door in two minutes and, if necessary, removed just as quickly. The main framework carries an aluminium hood wide enough to span two buckstays and as tall as a door. The vertical and bottom edges of the hood are fitted with plastic foam sealing strip and the roof carries flaps that can be adjusted to close off the channels in the buckstays.

Smoke and air, withdrawn from near the top of the hood to take advantage of their buoyancy, pass down a vertical duct and through a star shaped turbulence inducer which is also calibrated to serve as a flow meter. Cold air can enter the hood through three pairs of adjustable vents. When the machine is in position and the extraction rate (850 m^3/h) is set, the flow round the edges of the roof is examined. If smoke leaks outwards the gap between the roof of the hood and the battery wall is reduced and some of the vents are closed. If there is an obvious inward leakage some of the lower vents are opened. Either way the adjustments continue until the flow past the edge of the roof is negligible, i.e. the pressure across it is balanced. This procedure prevents loss of smoke, admits the maximum amount of cooling air to condense tars and yet minimises any change in the natural rate of leakage from the oven because the reduction of pressure within the hood is kept to the minimum needed to retain the smoke.

Samples are taken from the axis of the main duct through a sharp edged nozzle through which gas is drawn approximately isokinetically. Smoke particles and tar vapour are removed by two Greenberg-Smith impingers charged with stabilised trichlorethylene as shown in figure 5. The sampling rate is measured by a wet gas meter and both sample and mainstream temperatures are monitored so that all flow rates can be converted to a common basis. After use the solutions are filtered to remove insoluble material, evaporated to dryness and heated to 110°C for one hour to drive off water. The weight of the residue is taken as the tar content of the sample. If necessary the solutions can be analysed in detail but the simple "total tar" value is the most convenient basis for the general assessment of leakage rates.

Factors that affect the accuracy of such assessments include deposition of tar on the walls of the apparatus upstream of the sampling nozzle, incomplete recovery of tar from the sample stream, change in pressure drop across leaks and anisokinetic sampling. Weighing and cleaning representative areas of the hood showed that deposition was less than 1% of the total tar flow. Sample loss, which is caused by carry over of spray can also be kept to 1% or less by using two impingers in series. Change in pressure drop is minimised by the procedure described above but some slight suction within the hood is inevitable. Figure 6 compares the average suction at three levels in the hood with the pressure inside the oven and shows that, in effect, sampling depresses the base line by a factor ranging from 1.01 initially to two or more after thirteen hours. Since the leaks are all narrow, flow through them is laminar and will increase linearly with pressure drop. However, the effect of this on tar emission will be counteracted to some extent by the change in composition. Tar evolution decreases with time and will have become very small some hours before an oven is due for discharge. Therefore, although the induced increase in gas leakage rate becomes proportionately greater towards the end of a carbonising period there will not be a corresponding increase in tar emission. The effect of oversampling on the estimation of total emission from a works, i.e. the sum for ovens at all stages of carbonisation, is unlikely to exceed half that indicated by the ratio of the areas under the curve in figure 6, i.e. 3-4% rather than 7.5%. The anisokinetic sampling error was also small. Impinger resistance reduced the sampling rate from a nominal 10 l/m to 6.7 l/m so gas entered the sampling nozzle at 7 m/s compared with 12 m/s at the axis of the duct. Figure 7 shows the error calculated for a mainstream velocity of 12 m/s and its variation with particle size. For smoke sized particles all <5 μm and most <1 μm, the error resulting from the reduced sampling rate would be oversampling by about 1%.

To summarise

Deposition on hood etc	- 1%
Incomplete recovery	- 1%
Increased flow through leaks	+ 4%
Anisokinetic sampling	+ 1%

giving a net error of about +3%.

4. QUANTIFICATION OF VISUAL GRADES

The method described above was used to measure about ten doors in each grade at each works. Gradings were agreed by two observers before setting the hood in position and again after to ensure that the grade had not changed during sampling. The procedure was the same for all tests except for the duration which varied between about 5 minutes for grade 4 and 50 minutes for grade 0. Figure 8 shows the range and average values of emission rates for about ten doors in each grade. Despite the overlap the differences between adjacent grades are highly significant*. The total emission rate for each grade is the product of its mean rate $\bar{e}_g$ and the number of doors in the grade n_g. Total emission from all doors at a works is the sum of those for the grades

$$E = \sum_{g=0}^{g=4} \bar{e}_g n_g$$

Leakage is already monitored at all works by the "Agreed Uniform Technique" so it would be convenient if this scale could be quantified. CRE staff learnt the technique in collaboration with B.Ca.R.A. and, in three independent surveys, agreed to within ± 0.75% with each other and ± 0.8% with the experienced observers from B.Ca.R.A. Leaks were then graded on both scales and average emission rates for each grade were determined. The results are compared in figure 9 which shows that minor differences in grade limits have little effect on the emission rate curve and would not, of themselves, prevent either scale being used quantitatively. The crucial difference is in grade 4; not only is its bottom limit higher on the B.Ca.R.A. scale, but, more importantly, it includes all ignited leaks irrespective of size and these cannot be measured. Thus the higher bottom limit results in a higher average emission rate for those

---

1st series 5% level in Students t test

2nd series 2.5% level in Students t test

leaks that can be measured and the inclusion of ignited leaks attributes this high rate to leaks that are obviously emitting much less. Consequently the contribution of grade 4 doors and hence the total emission from the works is exaggerated.

Separate grading of leveller doors also complicates quantification because their emissions cannot easily be separated from those of the main doors. At the works tested this complication was minimal because leveller door leaks were few and small but this is not so at every works. The combination of grading separately leaks that cannot be isolated for measurement and including in the highest grade ignited leaks whose emission rates are trivial makes the Agreed Uniform Technique as currently defined unsuitable for quantification. However, this is no disparagement of a technique that was designed, not to measure emission, but to assess urgency of repair and which is therefore justified in regarding a small flame, with its potential for further damage, as more important than a larger but unignited leak.

Results from three different works show a three to fourfold range of emission rates corresponding to nominally identical visual grades, figure 10. One possible explanation is that rapid leakage through a small fault in a self sealing strip or slow diffusion through numerous cracks in a luted seal might alter the appearance of smoke formed by equal amounts of tar, but this cannot account for the difference between the two works with self sealing doors. Other possible causes include differences in feedstock, oven size - especially height, oven temperature, seal temperature, etc. In short the reason for the disparity shown in figure 10 is not yet known. The effort required to identify it could, however, outweigh any likely practical advantage. Instead it would be more effective to calibrate the smoke grades at individual works wherever quantitative emission rates are required. Meanwhile, although visual grading by the "Agreed Uniform Technique" provides a reliable method of monitoring changes within any one works it cannot be used with certainty to compare one works with another.

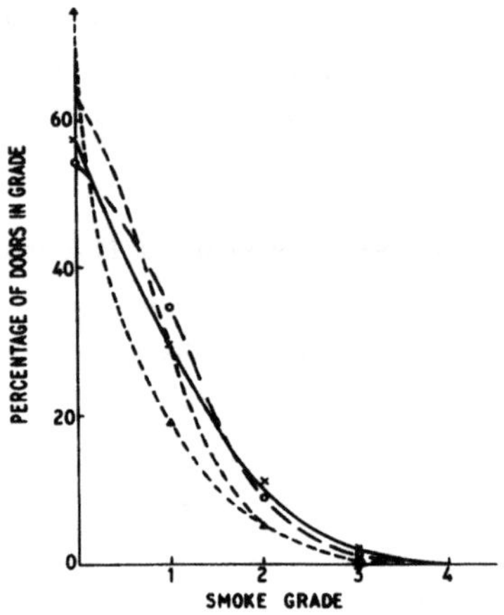

FIG.1 REPRODUCIBILITY OF VISUAL GRADING

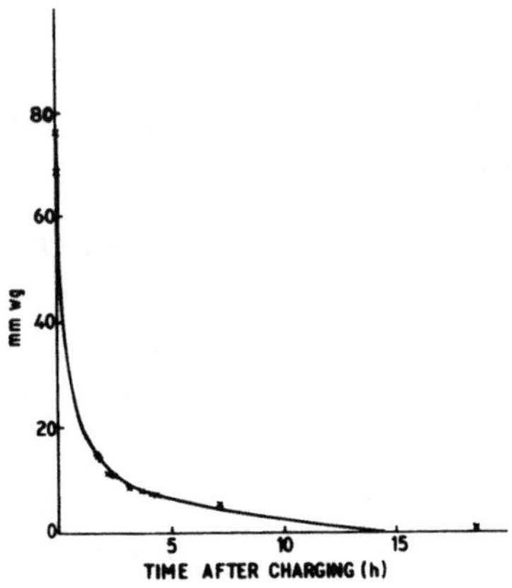

FIG.2 TYPICAL VARIATION OF PRESSURE DROP ACROSS LEAKS THROUGHOUT CARBONISATION.

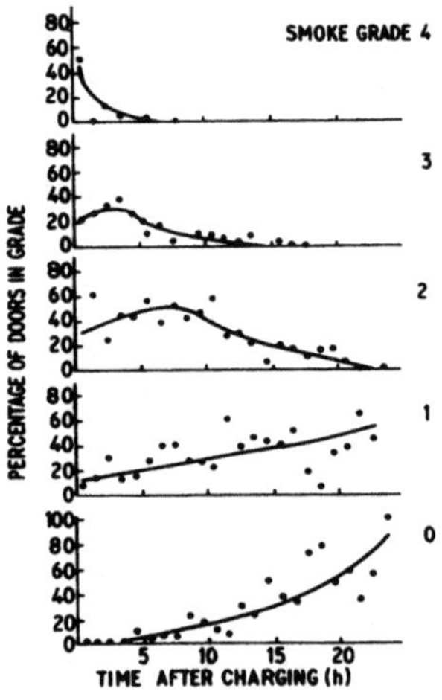

FIG.3 VARIATION OF SMOKE GRADES DURING CARBONISATION

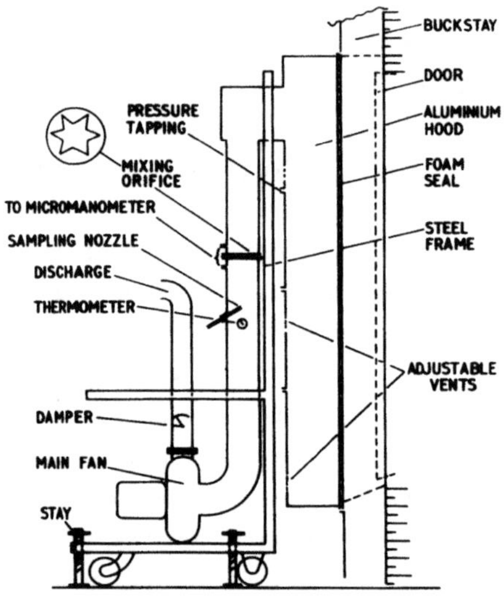

FIG.4 MOBILE APPARATUS FOR COLLECTING LEAKAGE FROM COKE OVEN DOORS

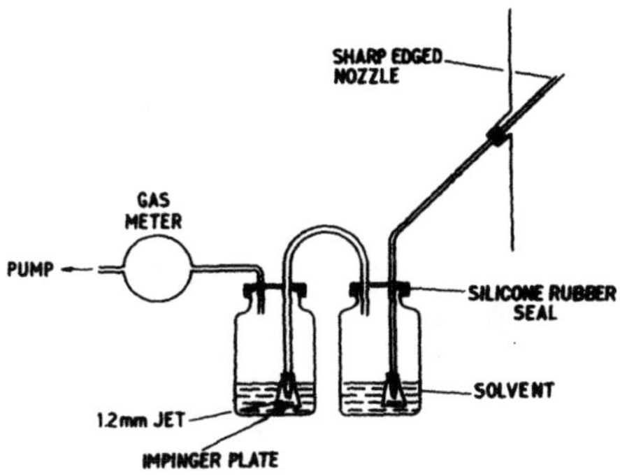

FIG.5 IMPINGER METHOD OF COLLECTING TARS

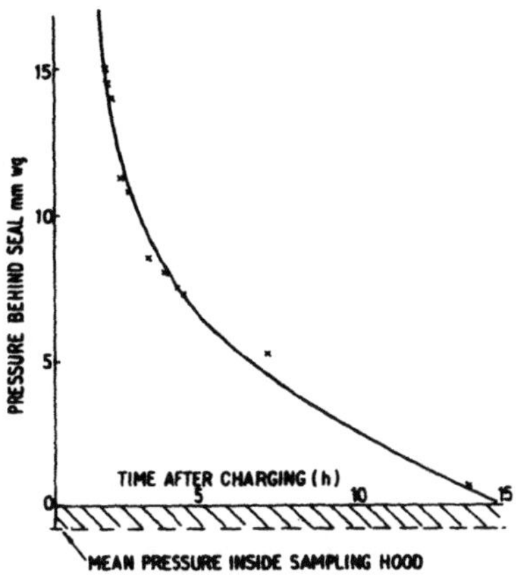

FIG.6 EFFECT OF SAMPLING APPARATUS ON PRESSURE DROP ACROSS DOOR LEAKS

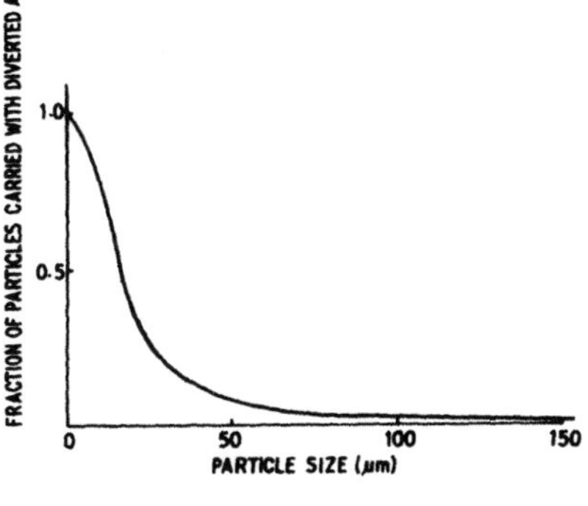

FIG.7 VARIATION OF ANISOKINETIC SAMPLING ERROR WITH PARTICLE SIZE FOR A MAINSTREAM VELOCITY OF 12m/s

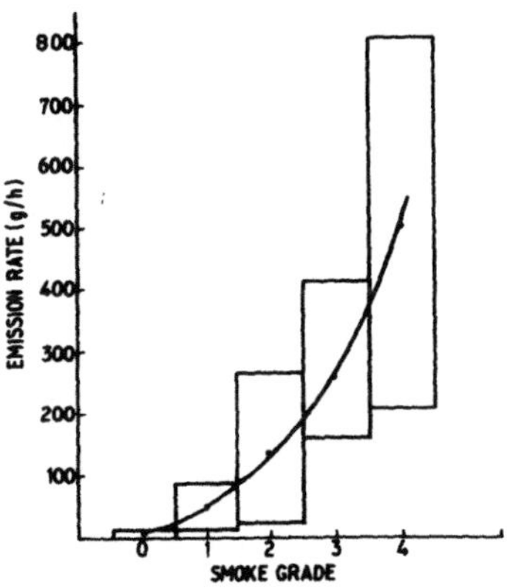

FIG.8 BSM EMISSION, RANGE AND AVERAGE, FOR EACH SMOKE GRADE AT ONE COKE WORKS

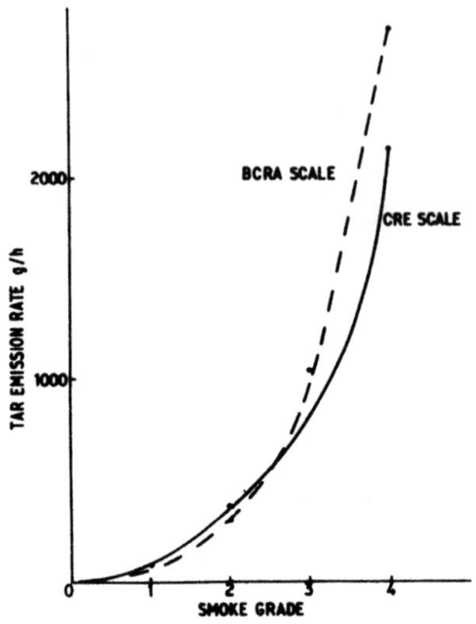

FIG.9 AVERAGE TAR EMISSION RATE FOR EACH GRADE AT ONE COKE WORKS

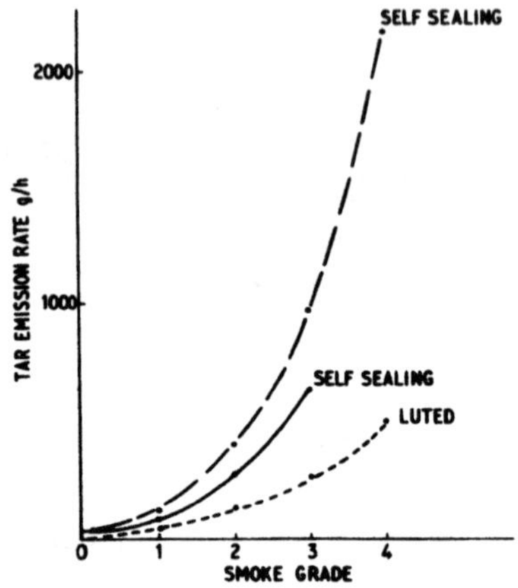

FIG.10 RELATIONSHIP BETWEEN EMISSION RATE AND SMOKE GRADE AT 3 COKE WORKS

SECOND TECHNICAL SESSION: COAL CHARGE PREPARATION AND UNCONVENTIONAL COKING TECHNIQUES

Chairman: M. BONAVENTURE, Directeur technique
Société Hainaut-Sambre

- Further development of the Precarbon process
  W. ROHDE, Bergbau-Forschung GmbH

- The manufacture of blast furnace coke in horizontal ovens from blends without normal quality characteristics by the combination of the preheating and stamp charging techniques
  J. VERZAUX, Engineer Houillères du Bassin de Lorraine
  T. ECHTERHOFF and H. FRICK, Directors, Kokereigesellschaft Saar mbH

- The use of pre-compacted coal for the manufacture of blast furnace coke
  R. VIDAL and R. MUNNIX, Centre de Recherches Métallurgiques Liège

- Formed coke from high volatile coal via char- the CdF process
  R. GUERIN, Charbonnages de France, R. DEGRAVE, Houillères du Bassin Nord et du Pas-de-Calais, and J. VERZAUX, Houillères du Bassin de Lorraine

FURTHER DEVELOPMENT OF THE PRECARBON PROCESS

W. Rohde

Bergbau-Forschung GmbH

Summary

The further development of the Precarbon process is designed to achieve rationalisation in operation and reduction of capital expenditure and running costs.

At the experimental coking plant, various technical measures were tested with the above end in view:

- A twin-stage flash drier, which, in a single column, dries the coal and preheats it to approx. 200°C.
- A single-stage flash drier with intermediate separation and dual product acceleration, which similarly heats the coal to approx. 200°C.
- A chain conveyor permanently connected to the chamber ovens of a coke battery.

Operational experience with the newly developed Precarbon equipment will be described; particular attention will be paid to the thermal characteristics of the preheating apparatus in comparison with the conventional double-stage flash drier.

Three different locking mechanisms were tested for the new charging conveyor. The tapered connection proved particularly suitable. For purposes of demonstration, a charging conveyor was connected up to four operating ovens. The shut-off devices in the transfer chutes are of the modified tapered connection type; they were subjected to continuous testing and emerged as the best.

As an introduction, I should like to briefly describe the Precarbon process for the thermal pretreatment of coal, as it has been successfully applied in four batteries with an overall capacity of 6 million tonnes of coal per annum. At all these plants, the economic advantage of this technique is based on the improvement of coke quality with a simultaneous increase in throughput, together with the reduction in feedstock costs due to the higher proportion of cheap coals with low coking capacities. The first Precarbon plant at the Gary Steelworks, USA, may be quoted by way of example: 57 large-capacity ovens have, for a long time now, been operating on our technique with about 106 oven-cycles per day - that represents a gross coking period of 12.9 hours. The ASTM stability factor of the coke is around 60%, corresponding to $M_{40} \approx 80$. The Precarbon coke's primary use is in large blast furnaces with 12.3 m hearth-casing diameter (Figure 1).

The Gary plant was to us both a confirmation of the preheating technique and at the same time an incentive to develop the Precarbon process further, to achieve lower capital expenditure and rationalisation of operation by means of improvements in apparatus and techniques. This paper sets out to describe the further development of the Precarbon process.

Since 1978 this development project, subsidised by the FRG's Ministry for Economic Affairs, has been the subject of research at the Prosper Experimental Coking Plant. Essentially it is characterised by two technical objectives:

A) Testing of a new twin-stage and a new single-stage flash drier
B) Testing of a new charging method.

With regard firstly to A) (Figure 2)

Although the previous double-stage flash drier proved itself to be an exceptionally effective unit with a high thermal efficiency, there are two ways in which it falls short of optimum energy use - the coal dried at the end of the first stage drops to the bottom of the second stage with no recycling of energy, and is then once more brought up to the height it was at previously. This transport of the coal requires energy in the form of electrical power. At the end of the first stage and at the end of the second stage, high-efficiency cyclones separate the coal from the heat-carrying gas. The pressure loss in the cyclones indicates the mechanical energy loss and has to be made up by the blower, which also uses electrical current.

The mechanical energy losses are halved in a new layout of the first and second stage (Figure 3). The coal is dried and preheated in a column with no intermediate separation. As in the conventional layout, the coal is first dried in the somewhat colder heat-carrying gas current; then - after mixing with hot combustion gases - preheating takes place to a temperature of approximately 200°C. The temperature profiles in the two adjacent flash drier columns and in the twin-stage flash drier column are virtually identical (Figures 4a and 4b).

The intensive heat exchange in the flash drier is typified by a steep temperature drop in the acceleration zone, i.e. after the coal has been fed in. The second acceleration zone as such is missing in the twin-stage flash drier pipe. Changing the flow conditions, however, gives rise, once heat has been applied, to a kind of second acceleration zone in the twin-stage flash drier as well. Be that as it may, the twin-stage drier is not identical with its predecessor in all points. Thus there are new specifications for the blower for example - the temperature of the gas to be propelled is substantially higher than in the conventional design, as it is the entry temperature of the heat-carrying gas at the foot of the twin-stage. However, this gas temperature is also identical with the waste-gas temperature; thus the waste-gas loss is higher than in the conventional design.

This should not be seen as an unmitigated drawback; as waste gas temperatures of less than 200°C entail a falling short of the sulphur dew point which means corrosion, if partially scrubbed coke-oven gas is underfired.

The following series of figures gives an impression of the test plant, which was designed for a throughput of 2 t/h and which was installed in the steel framework of the older 10 tonne plant (Figure 5).

Our attention was particularly drawn to the supply of the hot heat-carrying gases into the flash drier. The mixing zone must be designed in such a way that peak temperatures can be quickly lowered by an intensive intermingling and turbulence of the two gas currents; this means that overheating phenomena on the coal particles can be virtually eliminated. Figure 6 shows the principle of the mixing chamber: The combustion gas cooled by return-feed vapours is brought out of a combustion chamber tangentially to the vertical heat-carrying currents, i.e. under torsional force. The turbulence arising when the two flows are mixed is sufficient to immediately dissolve any streamers of hot combustion gas.

In conclusion it may be said that this twin-stage flash drier unit is operating faultlessly at the experimental coking plant and that it is fully living up to expectations.

In a further test the existing double-stage 10 t/hour plant was converted to a single-stage with dual acceleration. The schematic diagram at Figure 7 shows the operating principle.

Our design is distinguished from the conventional single-stage flash drier by a further acceleration stage half-way up - heat-carrying gas and solid are briefly separated by the centrifugal principle and then reunited. The solid is retarded and diverted, i.e. it receives a downward speed component. The renewed acceleration of the solid in the heat-carrying gas allows better exploitation of the fall of temperature still present at this height/level (Figure 8).

The temperature profiles measured in the flash drier illustrate this point (Figure 9).

In contrast to the above-described twin-stage principle, this is a single-stage operation, i.e. the wet coal comes in contact with relatively hot heat-carrying gas with a temperature of approximately 700-800°C as compared with only about 300°C in the twin-stage.

Both variants of the technique are characterised by waste-gas temperatures which must necessarily be substantially higher than the preheating temperature of the coal - in the first variant these are about 300°C, in the second approx. 250-280°C.

As already mentioned, Variant I requires this high waste-gas temperature as this is identical with the entry temperature of the drier stage (Figures 10a and 10b).

In summing up, the following points may be made:

The two new flash driers achieve gas/solid separation with just one single column and a single set of cyclones. Construction costs and hence capital outlay are thus considerably reduced; also the power of the blower can be decreased due to the reduced pressure loss. Set against these advantages there are the higher waste-gas temperatures and the fact that thermal efficiency is somewhat reduced. On the other hand, the higher waste-gas temperatures meet operational requirements for the avoidance of sulphur corrosion.

These new flash driers should, therefore, be considered when future Precarbon plants are designed.

In the second part of my paper I should like to go into the further developments of the charging system.

The previous distinguishing feature of the Precarbon charging technique was the arrangement of chain conveyors in a bridge layout above the oven roof and the connection between chain conveyor and chamber oven in the form of a travelling chute (Figure 11).

Of this system too, it can be said that it has proved itself in industrial conditions. The bridge layout has a not inconsiderable technical drawback however; this applies particularly if a cantilevered bridge construction on two supports is used, on the assumption of high calculated loads. The most recent operational Precarbon plant, Donner Hanna, already has a substantially lighter bridge construction as, in this case, multiple supports were provided for the chain conveyor bridge; however, we have formed the opinion that, while retaining the basic idea, the charging system should be still further simplified and that this constructional simplification would bring about operational advantages.

It has become clear, in the course of preheating applied on an industrial scale over many years, that any temporary return to wet coal gravity-charging is out of the question. One factor is high availability of the preheating system which has meant that no real case can be made for the alternative system with preheated and wet coal and the resultant high percentage stand-by time for machines and apparatus. The argument that the preheating system should also keep the traditional charging-car ready for operation on the oven roof is, therefore, irrelevant and the way is clear for a new layout of the chain conveyors.

The direct and permanent connection of the chain conveyor to the chamber ovens is particularly advantageous in many respects, as shown by the schematic diagram at Figure 12 and by the photograph of the demonstration plant at Figure 13.

This new charging technique offers the following advantages:

1) Between the chain conveyor or its discharge points and the chamber ovens there are no moving, telescopic connecting components. Consequently all mechanical and emission problems are obviated.
2) The support framework needed for the chain conveyors is minimal as compared with the bridge layout.
3) Chain conveyors and discharge points are at eye level and, therefore, easy to maintain.
4) The potential for the automation of the charging process is incomparably greater with the permanently connected chain conveyor.

On the basis of the development work that has been carried out, I can report the following:

Three separate outlets with different connection components were designed for a suitability test. An operational trial over several months demonstrated that the "bell-feed arrangement" is far superior to all other types of connection.

Contrary to expectations, temperatures in the oven crown do not cause any problems, so that the bell-feed can be inserted underneath the oven roof. Blockages do not occur underneath the bell-feed as, firstly, a cleaning blade removes coking residue adhering in the charging hole before charging takes place and, secondly, the degraphitisation phase helps to reduce ceiling graphite and coke.

Even if there is a lateral layout for the charging conveyor, inadequate charging profiles do not occur in the oven, provided an appropriate choice is made of preheating temperature and coal size consist, as can be seen from Figure 14.

Finally, Figure 15 is a photograph of the demonstration plant for the "new chain-conveyor charging technique".

In summary, it can be said that the Precarbon process had great potential for industrial development, which has led, in the recent past, to the developments described in this presentation. We are convinced that these new developments will lead to operational improvements and reductions in costs.

Figure 1. The Gary Precarbon plant, US Steel, Battery II.

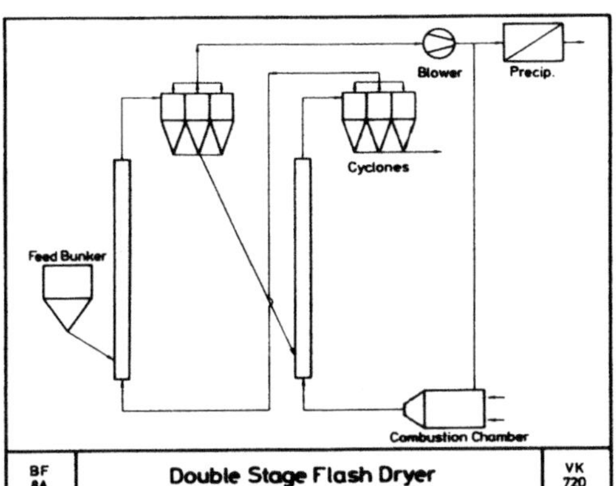

Figure 2. Diagram of double-stage flash drier.

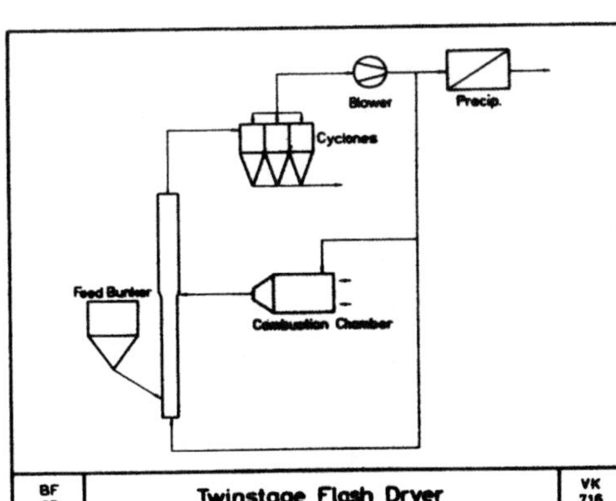

Figure 3. Diagram of twin-stage flash drier.

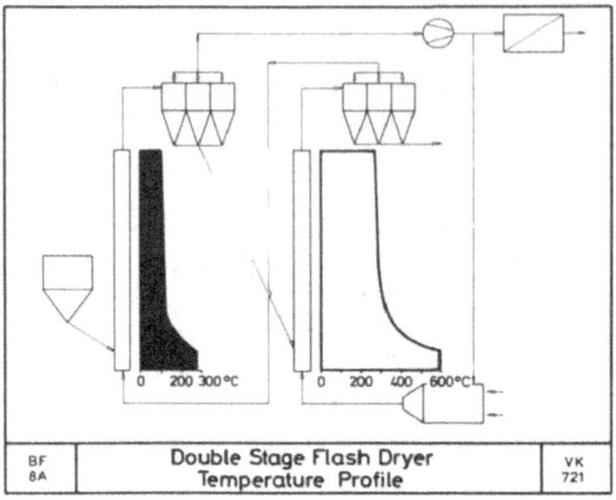

Figure 4a. Temperature profile for the double-stage flash drier.

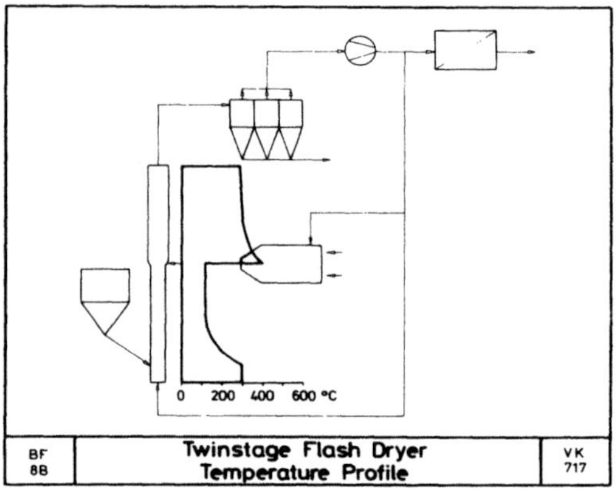

Figure 4b. Temperature profile for the twin-stage flash drier.

Figure 5. The twin-stage test rig.

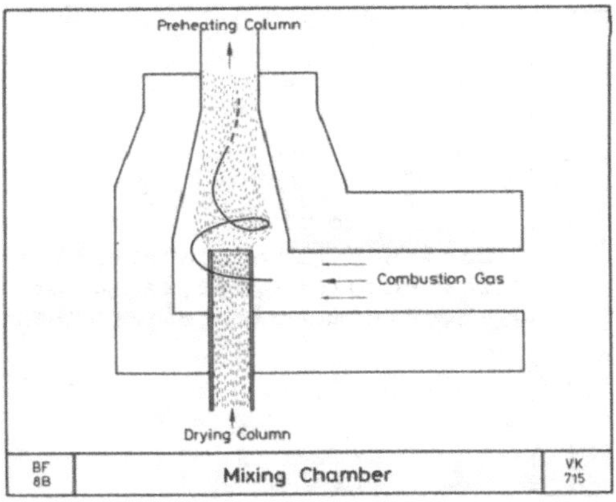

Figure 6. Diagram of the mixing chamber in twin-stage preheating.

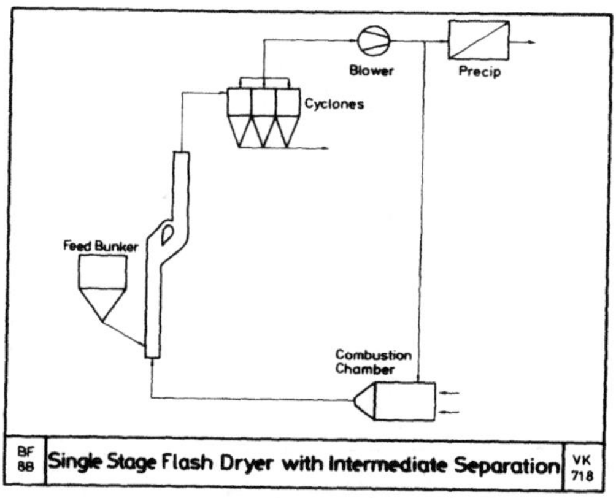

Figure 7. Daigram of single-stage flash drier with intermediate separation.

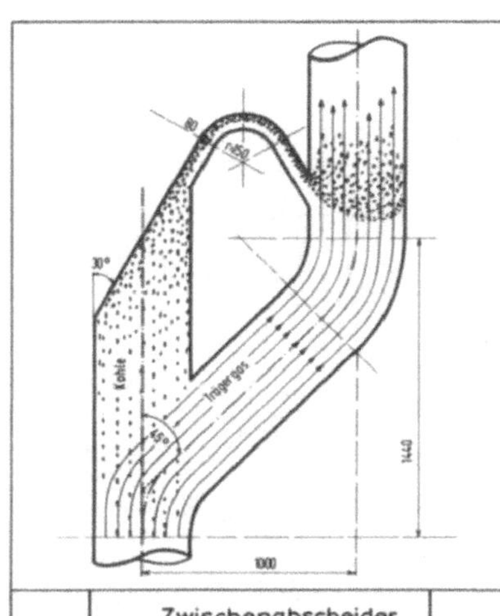

Figure 8. Diagrammatic representation of the second acceleration zone (intermediate separation)

Zwischenabscheider	Intermediate separator
Kohle	coal

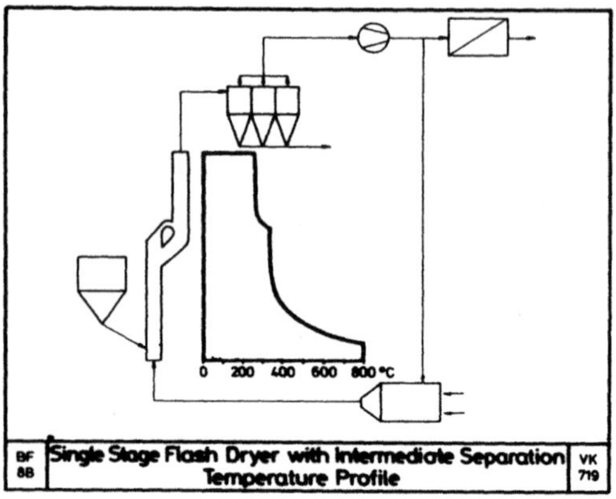

Figure 9. Temperature profile for single-stage flash drier with intermediate separation.

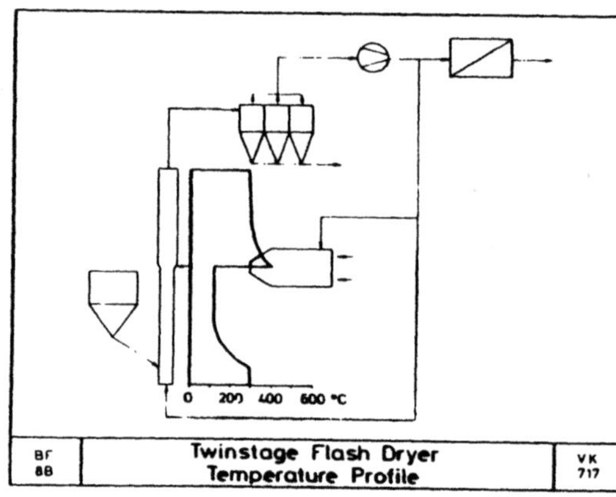

Figure 10a. Temperature profile for twin-stage flash drier.

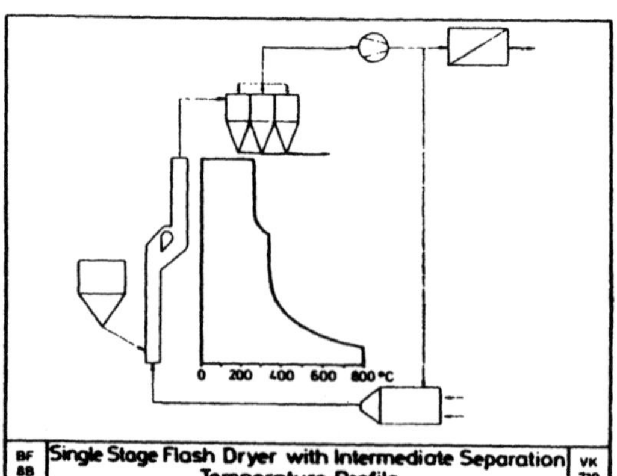

Figure 10b. Temperature profile for single-stage flash drier with intermediate separation.

Figure 11. Buggy and Redler.

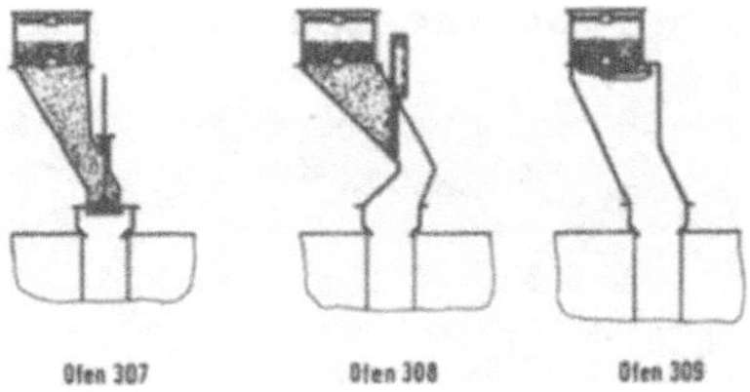

Figure 12. Schematic diagram of three couplings (connection between chain conveyor and oven chamber)
Ofen oven
Three different outlets for permanently connected chain conveyor.

Figure 13. The three couplings in prototype embodiment.

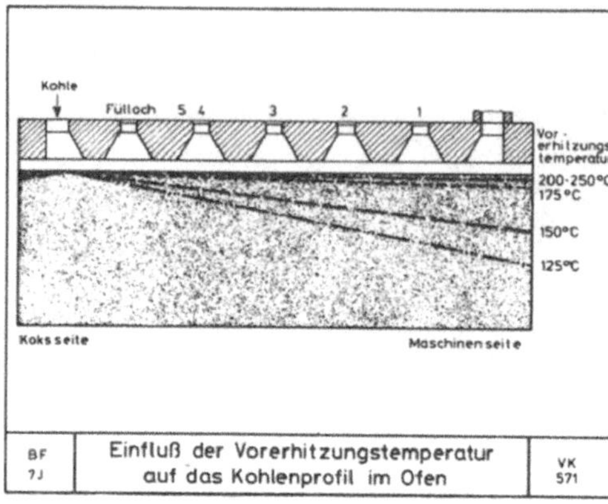

Figure 14. Influence of the preheating temperature on the coal level in the oven.

Figure 15. Demonstration unit for the new chain-conveyor charging technique.

THE MANUFACTURE OF BLAST FURNACE COKE IN HORIZONTAL OVENS FROM BLENDS WITHOUT NORMAL QUALITY CHARACTERISTICS BY THE COMBINATION OF THE PREHEATING AND STAMP CHARGING TECHNIQUES

J. Verzaux

Engineer in Chief at the Houillères du Bassin de Lorraine

J. Echterhoff and H. Frick

Directors of the Kokereigesellschaft Saar mbH

1. BACKGROUND

Over the past two decades, the trend towards increasing scarcity and expense of coking coals has become ever more marked. In some countries, the reaction has been to begin using ranks of coal less well suited to coking, although the quality often required for blast furnace coke sets limits to this trend.

With a view to diminishing the paramount importance of feedstock qualities for the quality of coke yielded, there has been keen competition to develop techniques allowing the limitation of the proportion of high-rank coals in the input blend.

Thus the stamp charging and preheating techniques came to be developed, particularly in the Lorraine and Saar coalfields which were seeking to maximise the use of indigenous coals for the production of blast furnace coke.

As the efficiency of these two techniques seemed not to be based on the same principle, the idea came about of combining the two processes, after a binder had been added.

2. TESTS AT EXPERIMENTAL STATION CARRIED OUT PRIOR TO THE CURRENT RESEARCH PROGRAMME

This technique was the subject of study and tests in 400 kg ovens, the research being initially carried out separately by the French and the Germans.

As early as 1960, the Marienau Experimental Station (SEM) showed that it was possible to produce a blast-furnace coke from a charge previously dried then compacted and exclusively made up of Lorraine coal and char.

The attached Table gives some of the most typical results of these tests (Tests 1, 2 and 3).

The findings were of such significance that the Charbonnages de France filed a patent (Patent No. 1 265 397 of 21 May 1960).

However, no development of the technique was envisaged at the time for technical reasons.

For its part the Kokereigesellschaft Saar (KGS) carried out, at its experimental station, preheating/compacting tests on blends unsuitable for the production of blast furnace coke by conventional methods. The results of these tests, given on the attached Table (Nos. 4,5,6,7), were also positive.

3. SEMI-INDUSTRIAL TESTS

HBL's Carling coking plant had available:

- a preheating installation which had served for the development of a system of steam-charging preheated coal (Coaltek system)
- an old stamp-charging battery whose closure was scheduled for the relatively near future.

HBL and KGS, therefore, decided to carry out joint experiments in the industrial ovens on stamp-charging of preheated blends with addition of an oil binder. A provisional test installation was constructed (see diagram of the installation), specially designed for the handling and introduction of the binder and the conveying of the blend into the stamping machine.

Previous tests had shown that at this stage in our experimentation, it was not possible to introduce the binder into the coal upstream of the crusher-preheater.

The installation was accordingly laid out in the following manner: the preheated coal in the CERCHAR crusher-preheater at a temperature of 170-180°C is stocked in a loading hopper; from this hopper it flows into a twin-screw mixer to which is added an oil bitumen heated to 150°C, to act as a binder. On discharge from the mixer (throughput 10 t/h), the blend is removed by 2 successive conveyors, directly to the stamping box.

By compacting these blends comprising 6% bitumen, it was possible to produce cakes 2.6 metres high with sufficient stability to allow problem-free charging. After compacting, the temperature of the cakes was still 150-160°C.

Nineteen different blends, based on both Lorraine and Saar coal, were charged. In total, there were 113 charges of 13 t.

The attached Table gives the results for the most significant blends. These results were similar to those obtained at the experimental station, thus showing that the technique permits the composition of blends of coal with high and low VM contents, both of which have only a low coking capacity (Nos. 8,9,10,11).

Furthermore, the reduction in residence time obtained for these charges in a battery otherwise operating on the wet-stamping principle, together with the increased density of the charge when dry, showed that the increase in productivity was similar to that of preheated charging techniques.

Moreover, the problem of carry-over and of the charge's lack of homogeneity (thence unevenness of heating) does not occur with the preheating/stamping process, as the total charge reaches the oven as a large slab with a high degree of compactness and homogeneity.

With regard to the nature of the blends and some elements of their preparation, it is important to emphasize that:

- preheating/stamp-charging is a special technique calling for specific blends,
- the constituents of these blends and the additives (thinners,

etc.) come from a much wider range than in other techniques,
- coke dust cannot be used on its own as a thinner,
- the VM content of the blends must not exceed a threshold value (of the order of 30%),
- the qualities of the coals (plastic properties in particular) is less crucial than in other techniques,
- the degree of fineness of the coke paste is of the utmost importance.

4. INDUSTRIAL TESTS

The success of the initial tests at the experimental station and on an industrial scale ought not to give the impression that the technique has been fully and finally developed for industrial use. Considerable research efforts are currently in train at laboratory and experimental station level as well as at industrial development level.

Table 1

Composition of the blend	PRV	Tests at experimental station			KGS tests in 1977				Industrial tests at Carling I (December 1977 - January 1978)			
No. of the blend		1	2	3	4	5	6	7	8	9	10	11
Long-flame gas coal A	0.8	36.2	71.4	-	54	36	18	-	54	51	36.7	36.7
High-volatile bituminous coal	0.75	-	-	-	-	18	36	54	-	-	-	-
High-volatile bituminous A	0.9	38.1	-	76.2	-	-	-	-	-	-	17.9	17.9
Medium-volatile bituminous 301	1.05	-	-	-	-	-	-	-	-	10	-	-
Low-volatile bituminous 203	1.5	-	-	-	15	15	15	15	-	-	15	15
Low-volatile bituminous 202	1.8	-	-	-	-	-	-	-	15	-	-	-
Char 500°C (20% VM)	-	12	14.9	10	-	-	-	-	-	-	-	-
Delayed coke	-	-	-	-	20	20	20	20	20	27	15	19.7
PK	-	9	9	9.1	5	5	5	5	5	6	9.4	4.7
Binder	-	4.7	4.7	4.7	6	6	6	6	6	6	6	6
M 40	-	80	78	79.5	75.9	78.6	76.8	76	77.1	83.0	83.8	79.6
M 10	-	7.8	7.7	7.5	8.8	8.2	8.4	8	8.2	8.6	7.9	7.6
I 40	-	-	-	-	-	-	-	-	52.2	63	64.6	54
I 20	-	-	-	-	78.7	78.5	79.3	77.8	79.1	78	76.2	82.2
I 10	-	-	-	-	19.7	19.8	18.7	20.3	18.8	19.4	22.1	14.5

Diagram No. 1

December 1977 to February 1978

Installations built at Carling I for the combined pre-heating/stamp-charging technique

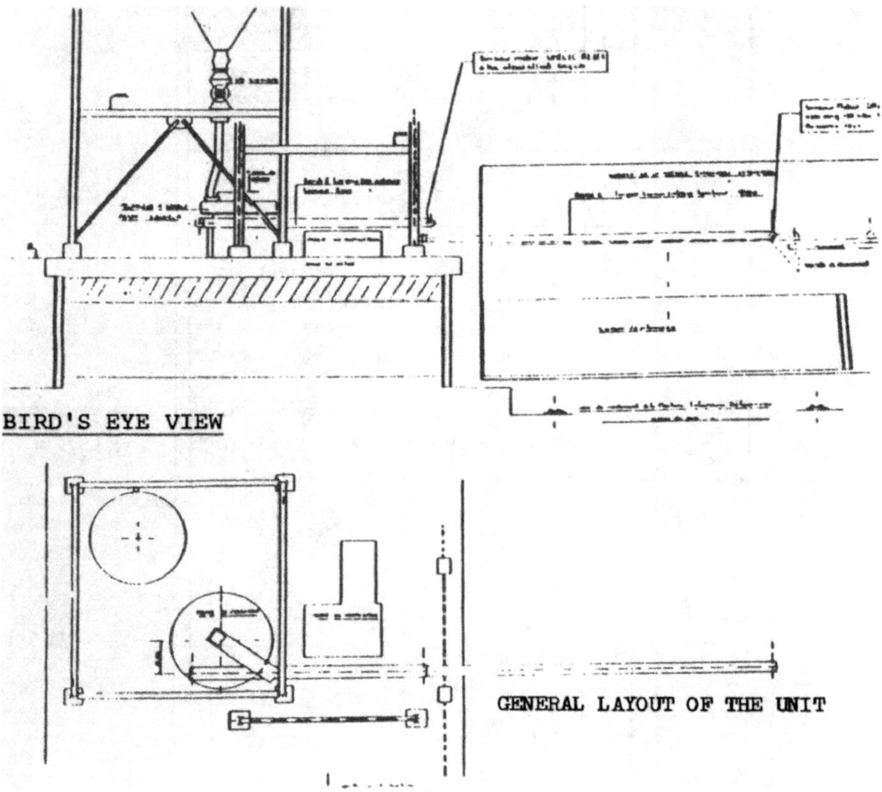

PART II

The positive results obtained in the above-described tests prompted us to embark on a research project designed to develop the preheating/binding/stamp-charging technology of coke production to industrial scale.

We looked for a drying system suited to this combination of techniques and our choice fell on the conventional rotating flighted drying drum and direct drying by means of contact with a drying medium. In order to test whether this system was appropriate for our purposes, tests were arranged with 5 manufacturing firms.

These tests were carried out on 3 drum systems and a "rapid drier" with scoop-type paddles, in each case with a direct-contact drying medium, and also a drum heated indirectly via a pipe assembly. The processes using direct-contact waste gas seemed unsuitable as they involved combustion techniques which meant that the gases were propelled at low pressure and were not very dense. This, say specialist firms, makes it difficult for a drying drum to achieve an oxygen content of $<8\%$ in the drying medium. From the safety point of view, this is undesirable given a residual water content of 0.5% and a target coal temperature of 150-200°C. Moreover, to maintain prescribed threshold values, such systems require much measuring and adjustment, which experience has shown to reduce plant availability. Quite apart from the safety aspects, as low an oxygen content as possible in the drying medium is desirable from the point of view of the coal itself, as, with the prevailing temperature, the oxygen may oxidise the coal and cause deterioration in its quality. The indirectly heated drum system was eliminated as a proportion of the coal baked on to the heating pipes and carbonized.

Our test results led us to choose a technique involving the inert drying medium, steam, in direct heat transfer. After preliminary tests carried out in conjunction with a specialist firm, the system was tailored to our requirements and design was completed. A 1:1 model was used to optimise the geometry and layout of the paddles inside the drum to achieve the best possible distribution of coal over the cross-section. A test installation with a nominal capacity of 8 t/h was ordered and constructed last year at Saarbergwerke's Fürstenhausen coking plant. A brief description of its structure and function is set out below:

The nucleus of the installation is a rotating drying drum with paddles inside, through which the finely distributed coal passes and is simultaneously dried by superheated steam with which it is in direct contact. The steam originates from the coal itself. It is fed into the cycle under pressure through a fan, scrubbed in a gauze filter and superheated in a gas-heated recuperator. Any surplus of steam is scrubbed and vented off, under regulated pressure, to a flue. A condenser operating indirectly in the cycle can deal with any heat surpluses in the start-up and cut-off phase. The inert steam cycle, which has to be built up at start-up before the coal is fed in, is created by the evaporation of finely ducted water. To comply with an official regulation, the O_2 content generated must be less than 2%. The dried coal is collected in a bunker, and is later weighed charge by charge and fed into a positive mixer with a prescribed quantity of binder. Mixing was carried out charge by charge as this made it possible to achieve the desired quantities more precisely than was the case with continuous operation. The diagram at Figure 2 shows the materials flow and the operating principle of the system.

The plant was completed in the middle of 1980 and was then commissioned. At the beginning, minor defects were spotted and eliminated. These were mainly leaks at sealing components and flanges which were noticed because steam escaped, due to the plant being operated under pressure.

Also, to avoid condensation phenomena, some parts of the plant were given extra insulation. But even the earliest tests demonstrated that the coal drying system was living up to expectations and the individual components were fulfilling their role. The mixture zone downstream, for the addition of a prescribed amount of binding agents, is operating successfully so that, on leaving the mixer, the preheated coal can be transported in the open without giving off dust or causing any hazard. The venting-off of the surplus steam does not give rise to any emission problems, as the gauze filter incorporated in the vapour circuit scrubs the vapours thoroughly.

A series of drying tests were carried out with different parameters to establish key data for the plant. Coal blends were dried with a size consist of:

$>$ 10 mm	0%
$>$ 3 mm	5%
$<$ 0.5 mm	50%

The water content of the coal is between 9 and 16% depending on the blend. The key data for a given blend and operating method are given in Table 2.

The striking features are the low temperature level at which drying is carried out and a favourable specific heat consumption. Preliminary tests at the semi-industrial stage showed that, given suitable binders, low coal temperatures lead to no disadvantages in cake stability and the coke quality obtained as compared with a high coal temperature. It remains to be determined in the course of future tests to what extent the specific heat consumption in preheating and the throughput of the coke oven battery are affected by coal temperature.

The low Δ-T between steam and coal at the drum outlet demonstrates that the heat transfer from the drying medium to the coal must be very intensive. We believe that the drying process takes place as follows:

As a uniflow system is used, the superheated steam comes in contact with the cold coal and, by condensation on the coal particles, raises the water content to the limit of absorption power, before a temperature of 100°C is reached. The subsequent drying and heating of the coal is done very gently, as the absorbed water must be condensed before the application of the sensible heat. Therefore, the coal particles are not subject to overheating during the drying and preheating process. That is important as higher temperatures can impair the coal's quality.

The plant has since been operating in a systematic working cycle for the manufacture of preheated, bonded coal. Special equipment conveys the coal into a normal operational stamping machine at the coking plant, where it is compacted by drop hammers to a large cake weighing about 21 t; it is subsequently pushed into a coke oven and carbonised. Occasionally larger amounts of coal (75 t) were prepared in advance and, until required for use, were left in the open overnight at temperatures of below $0^{o}C$. The temperature loss in the stockpiled coal was only about $10^{o}C$ after 24 hours and subsequent handling during transport and in the compacting machine tended to be better than when freshly preheated coal was treated.

Apart from operational testing of the process, blends with low coking capacities were also carbonised. This brought confirmation of the finding from the preliminary tests that this technique is particularly successful in improving the mechanical properties of coke when the input blends used have high proportions of weakly caking coals. Some test data are reproduced in Table 3 by way of example.

One portion of the blend in question was wet-stamped, whilst the other portion was preheated, bonded and coked in compacted form. It was established that with the very weakly caking blend (swelling index 1.5) the wet method cannot produce any usable coke, whereas the preheated, compacted blend yields a coke with good mechanical characteristics.

5. SUMMARY

Tests in the laboratory and in a pilot plant at Carlingen have shown that with the combined preheating/stamp-charging technique, blends with very poor coking capacity can be used and can yield good coke.

This prompted us to develop the combined preheating/

stamp-charging technique to the fully industrial stage.

In the context of this development work, a new coal preheating system was constructed at Fürstenhausen coking plant. In this system the coal is gently and safely dried and preheated at a low temperature level and with economic power consumption. The plant allows a coal throughput of 8-12 t/h. The coal preheated in this plant is compacted in a stamping machine located in the normal working cycle and carbonised in the existing coke ovens which are 4 m high. Early results have confirmed the findings from the laboratory tests and the field test at Carlingen.

Apart from the optimisation of coal preheating in conjunction with suitable binders, tests will be carried out in the near future on the behaviour during bunkering and transport of the coal/binder mixture. So far as the stamping machine is concerned, the intention is to minimise the amount of time and power used in compaction in terms of coal temperature and binder type bearing in mind that the cake of coal must be adequately consolidated. At the coke ovens, the characteristic coking parameters will be established and optimised.

In a further stage of development, the coking trials at Fürstenhausen coking plant will be transferred to the 6 m high coke ovens. At a third stage, the combined preheating/stamp-charging technique will be applied in continuous service in a battery at Carlingen.

The further development of the combined technique is being subsidised by the Commission of the European Communities. We should like to express our gratitude for this subsidy.

Table 2

Key data of a drying test

Water content of the coal	9%
Dryer throughput	9 t/h
Temperature of the steam at the drum inlet	400°C
Temperature of the steam at the drum outlet	165°C
Temperature of the coal at the drum outlet	150°C
Temperature of the steam in front of the filter	125°C
Temperature of the steam before entry into the recuperator	120°C
Temperature of the mixture of waste burner gas and surplus steam at flue inlet	160°C
Pressure at the drum outlet	25 mm water column
Residual water content of the dry coal	0.5%
Oxygen content in the vapour circuit	0.7%
Heat consumption	100 kcal/kg coal f

Table 3

PREHEATING/STAMP-CHARGING

Carbonization of a weakly caking blend

1. Wet stamped, carbonized in the test unit
2. Preheated, bonded, stamp-charged, carbonized in normal working cycle.

Input, conditions and results

Blends (wf) %		1	2
Saar high-bituminous coal	621	48.0	45.1
Saar fat coal	635	15.0	14.1
Lean coal	322	21.0	19.8
Lean coal	311	10.0	9.4
Coke dust		6.0	5.6
Binder		-	6.0
Water content		10.9	0.5
Ash (wf)		7.6	7.5
Volàtile matter (waf)		28.8	31.8
Swelling index		1.5	1.5
Dilatometer test			
Fusion		386	346
Reconsolidation		485	422
Contraction		26	37
Dilation		-26	-37
Strength and abrasion			
M_{40}		66.6	75.0
M_{10}		21.8	7.0
I_{20}		54.6	77.1
I_{10}		44.0	20.5

THE USE OF PRE-COMPACTED COAL FOR THE MANUFACTURE OF BLAST FURNACE COKE

R. Vidal and R. Munnix

Metallurgical Research Centre (C.R.M.), Liège - Belgium

Summary

The gradual exhaustion of the world's reserves of good coking coal necessitates a widening of the range of coals which can be turned to account for coking purposes.

Partial briquetting enables a good blast furnace coke to be manufactured in conventional ovens, from blends containing a high percentage of reputedly poorly-coking or non-coking coals.

Tests at a pilot plant have enabled us to study the parameters affecting this technique.

Coking tests on an industrial scale have shown that the introduction of 33% non-coking coal briquettes into the coal charge presented no technical problems.

A long-term blast furnace test showed that the new type of coke thus manufactured behaves in just the same way as a good conventional coking coal.

1. INTRODUCTION

For a number of years now the C.R.M. has paid particular attention to the development of new coking techniques which would make possible the consumption of poorly-coking coal whilst still producing a good blast furnace coke at a competitive price.

More particularly, we have conducted a large number of blast furnace charging tests on different types of formed coke ((1) to (3)) which have demonstrated that some of these cokes could indeed be used as a partial replacement for conventional coke. As yet, however, these new techniques have not reached a sufficiently advanced stage of development to cope with all the problems involved. In addition, the production of large quantities of formed coke would mean the replacement of a large number of existing plants and this would involve high investment costs.

From 1972 onwards we have been conducting research work oriented towards determining how, and to what extent, poorly- or even non-coking coal (according to conventional criteria) could be used in conventional coke ovens.

An in-depth examination of up-and-coming and present-day coking techniques led us to develop the BRICOKE process which consists of drying and crushing the poorly- or non-coking coal, adding an appropriate binder/solvent and briquetting the resultant blend under high pressure. The briquettes thus formed are then mixed to a good coal charge and the resultant mixture is then charged in classical coke ovens.

As the coke produced by this process was of a completely new type, it was necessary to provide a detailed evaluation of the coke based not only on its size consist and mechanical strength, but principally on the results of long-term blast furnace tests.

With EEC financial aid our initial task was to study at the Pilot Plant the various parameters likely to affect the quality of the coke obtained by partial briquetting. We then carried out industrial-scale trials which involved on the one hand the production of 25 000 tonnes of this new type of coke from a blend containing 33% non-coking coal briquettes and, on the other hand, charging 100% of this coke in a blast furnace with a hearth diameter of 8.2 m.

2. PRINCIPLES OF THE BRICOKE PROCESS WHICH ALLOWS THE CONSUMPTION OF HIGH PROPORTIONS OF NON-COKING COAL

The development of the BRICOKE process was based on the following facts:

1. Compaction of coal smalls improves the coking properties of coal with a high volatile matter content, as demonstrated by the stamp-charging technique which has been successfully applied for a number of years in the Saar, in Lorraine and in Poland (4);

2. Pre-heating of poorly-coking coal brings about a considerable improvement in the quality of the coke, but drying of the coal alone is responsible for much of this improvement (5,6,7);

3. The addition of suitable organic substances such as tar, pitch, asphalt and heavy oils greatly improves the coking properties of a wide range of coals.

The BRICOKE process developed by C.R.M. and its associates combines the beneficial effects of drying, the addition of a binder/solvent and compaction in order to turn a wide range of reputedly poorly-coking or non-coking coals into good coking coals.

Similar processes have also been developed in Japan (8), (9).

3. PILOT PLANT TRIALS

At the Pilot Plant the trials were conducted in the SIDMAR and C.R.M. experimental coke ovens and the briquettes were manufactured by INIEX.

Numerous tests were carried out aimed at determining both the effects of parameters affecting the quality of coke obtained by pre-compaction and defining the range of coals which could be turned to account.

In this report, which has been highly condensed, we will simply set out several examples of the parameters studied.

3.1. Effects of the binder/solvent, the pre-compaction and the percentage of briquettes in the coal charge

Tests were carried out on various types of coal in the International Coal Classification System (Ranks 200, 511, 621, 622, 623) but here, by way of example, we shall only be dealing with the results obtained with poorly-coking coal of rank 621 (Table 1).

After drying and crushing to a suitable size consist the non- (or poorly-) coking coal was briquetted using 6% pitch. The briquettes thus obtained were then incorporated in increasing proportions into a good basic blend (BB in Table 1).

Figure 1 shows the characteristics of the different types of coke obtained by incorporating increasing proportions of coal of rank 621 - in the form of smalls, smalls with added pitch and briquettes, respectively - into the basic blend.

We noted the following points:

1. Coal of rank 621 is definitely non-coking coal as incorporation of increasing proportions of this coal in the form of smalls into the charge substantially lowers the quality of the coke;

2. The addition of 6% pitch of the non-coking coal smalls reduces this degradation somewhat;

3. Pre-compaction together with the addition of pitch completely eliminates this degradation and even produces coke of a slightly higher quality than that obtained from the basic blend alone.

Then again, our tests showed that the charging density of the coal charge is dependent upon the quantity of briquettes incorporated, the maximum density being reached in the case of approximately 60% briquettes. These observations are very much in keeping with the results published by other researchers (4), (8).

3.2. The effect of the form of introduction of the poorly- or non-coking coal

Bearing in mind the flexibility of the process, it is possible to incorporate the poorly- or non-coking coal with the added binder/solvent into the coal charge in different forms, such as:

- conventional type briquettes (dry coal plus binder/solvent, followed by compaction);
- recrushed conventional briquettes;
- dry coal smalls coated with binder/solvent by means of a hot (90°C) mixing process;
- dry coal smalls plus solid binder/solvent.

In order to demonstrate the effect of the form of introduction of the poorly- or non-coking coal on coke quality we expressly incorporated into the conventional basic blend 40% of a mixture made up of 50% coal (rank 200) and 50% coal (rank 535) in each of the following forms in turn:

- dried smalls plus 8% solid pitch added;
- dried smalls coated with 8% pitch by a hot mixing process;
- briquettes bound with 8% pitch;
- briquettes bound with 8% pitch and finely re-ground (90% $<$ 2 mm).

The main characteristics of coal of Ranks 200 and 535 are summarized in Table 1.

The IRSID and MICUM strength indices of the coke obtained are shown in Figure 2.

It may be observed that the form of introduction of the coal has an important effect on the quality of coke produced:

- hot coating of poorly-coking coal substantially improves the quality of the coke as compared with charging the coal smalls and solid pitch;
- charging briquettes further improves the abrasive strength indices I20, I10 and M10, but lowers the cracking strength

index M40;

- the introduction of crushed briquettes brings about a slight deterioration of the abrasive strength indices I10 and M10 (as compared with the charging of whole briquettes) but greatly improves the cracking strength index M40.

This example shows that it is possible, by combining different forms of introduction, to adapt, within certain limits, the quality of the coke to meet the consumer's requirements.

4. INDUSTRIAL COKE OVEN AND BLAST FURNACE TESTS

The industrial-scale tests were carried out at the COCKERILL coke works, Flémalle and in COCKERILL blast furnace 5, Seraing.

4.1. Preliminary tests

With a view to confirming the results obtained in the experimental oven and examining any possible technical problems arising from the handling or separating of the briquettes in an industrial plant, it was decided to carry out a preliminary test involving the production of 2500 tonnes of coke by charging 40% non-coking coal briquettes (Rank 621) in a good basic blend.

Excellent results were achieved at the coking plant, as can be seen from the following figures:

I_{10} = 17.1 ; I_{20} = 79.4 coke fraction $<$ 20 mm = 1.6%

A charging trial in a small blast furnace with these 2500 tonnes of coke showed that operation of the furnace was in no way affected by this new type of product.

4.2. Long-term tests

The promising results achieved in the preliminary tests led the operators to conduct a long-term trial during which the two coke oven batteries were supplied with a coal charge containing 67% conventional basic blend and 33% briquettes containing 94% non-coking coal (Rank 621) and 6% coal tar pitch.

The production capacity of the briquetting plant and

especially the drier were distinctly lower than the coking plant's requirements and unfortunately the coal could not be dried prior to briquetting. This had a slightly adverse effect on the quality of the coke. Table II summarizes the characteristics of the coke obtained both during the test period proper and the reference period when only the basic blend was charged. It can be seen that apart from the effect on the M40 index, the characteristics of the different cokes are very similar.

With the entire COCKERILL, Flémalle coke works engaged on the production of the coke required for the tests, we were in an ideal situation to build up a stock of carbonization products, both during the production of the test coke proper and during the production of the reference coke.

In short, we noted that the introduction of 33% non-coking coal briquettes (Rank 621):

- does not affect the total, dry coke output but slightly reduces the output of blast furnace coke in favour of small coke;
- increases the gas output slightly along with the calorific value;
- does not alter the ammonia or benzol yield;
- increases the tar and benzol yields.

The 25 000 tonnes of coke produced by incorporating non-coling coal briquettes were charged in blast furnace 5 of COCKERILL-Seraing E, which has a hearth diameter of 8.2 m and an operating volume of 1108 m^3.

Table III summarizes the main operating results achieved during both the test period and the corresponding reference period where the basic blend only was charged.

The blast rate and the cast iron production remained unchanged. Then again, the coke rate and the quality of the cast iron improved as compared with the charging of normal coke.

We can, therefore, appreciate that the two types of coke

with similar IRSID strength indices and similar granulometric curves, but with a M40 index which drops from 82 to 76, behave identically within the blast furnace. It should, however, be noted that the granulometric curve of the new coke was more compact and that the average diameter of this coke remained at around 57 mm.

Whatever the case, it would appear that on the basis of these results partial-briquetting enables us to produce a completely acceptable blast furnace coke from blends containing reputedly poorly- or non-coking coal.

REFERENCES

(1) A. POOS, C. VANOSMAEL, N. PONGHIS.
Effect of coke sizing on blast furnace operation; special reference to formed coke.
Ironmaking Proceedings of AIME - 1969 - pp. 448-467.

(2) W. HERMANN, R. SCHMIT.
The Ancit Process. Agglomeration.
Proceedings of the Second Int. Symposium on Agglomeration. Atlanta, March 1977.

(3) C. PESCATORE.
The Ancit Process.
8th Regular Meeting of Committee on Technology of I.I.S.I. March-April 1976.

(4) K.H. FLASCHE, H. SCHMAUCH.
Erhohung der Schuttdichte der Einsatzkohle durch Briket-tierung.
Stahl u. Eisen, N2. 26, 29 Dezember 1977, p. 1325.
Increasing the bulk density of the coal charge by bri-quetting.

(5) D. HABERMEHL, G. LASK, H. PETAK.
Résultats récents concernant la production du coke à partir du procédé Precarbon pour le préchauffage des charbons.
Commission des Communautés Européennes.
Journées d'Information sur les techniques des cokeries, Luxembourg, mai 1975, p. 231.
New results on coke production from preheated coal using the Precarbon Process.
Commission of the European Communities.
Information Symposium on Coke Oven Techniques, Luxembourg, May 1975, p. 231.

(6) G.P. GRAHAM and V.J. PATER.
Etudes et développement du préchauffage des charbons au Royaume Uni, Tome 4, p. 203.
Research and development on coal preheating techniques in the United Kingdom, Volume 4, p. 203.

(7) G. PRUDHON and R. MARCELLINI.
Enfournement préchauffé. Etude systématique de l'influ-ence de divers paramètres.
Preheated charging. A systematic study of the influence of various parameters.
Commission des Communautés Européennes.
Journées d'Information sur les techniques des cokeries, Luxembourg mai 1975, p. 205.
Commission of the European Communities.
Information Symposium on Coke Oven Techniques, Luxem-bourg, May 1975, p. 205.

(8) M. YOSHINAGA et al.,
Present status of production procedure for briquette blend coke.
Nippon Steel Technical Report Overseas, No. 10, November 1977.

(9) T. IKECHIMA.
Reduction of coke rate and new coking process using non-coking coal.
I.I.S.I. 1976, Report of Proceeding.
Tenth Annual Conference, Osaka, October 1976.

TABLE I

MAIN COAL CHARACTERISTICS

Coal	Basic blend	Non-coking coal		Make-up coal
Reference	B.B.	-	-	-
International classification No.	434	621	200	535
Proximate analysis				
V.M./dry (%)	26.2	31.3	9.3	30.9
V.M./dry and pure (%)	27.7	34.6	10.4	32.4
Ash (%)	5.4	9.6	10.2	4.7
Sulphur (%)	0.81	0.59	0.90	0.60
Swelling Index	8.25	3.5	0	7.5
Arnu test				
Max. contraction (%)	27	20	0	30
Max. expansion (%)	81	-20	0	191
Total swelling (%)	108	0	0	221
Max. fluidity - Gieseler (dial division per minute)	336	19	0	5.400
Mean max. reflectance (%)	1.25	0.94	2.49	1.08
Maceral analysis				
Vitrinite (%)	58.3	46.9	46.2	68.1
Exinite • (%)	5.0	13.6	0	8.9
Inertinite (%)	33.8	34.8	42.6	19.2
Minerals (%)	2.9	4.7	11.2	3.8

TABLE II

CHARACTERISTICS OF THE COKE PRODUCED AT COCKERILL-FLEMALLE

		Reference period	Test period proper
Composition of the coal charge			
Basic blend	(%)	100	67
Coal (Rank 621) briquettes	(%)	0	33
Coke characteristics			
Moisture	(%)	3.37	4.12
Ash	(%)	8.56	9.04
Volatile matter	(%)	1.04	1.19
Screen analysis			
80 mm	(%)	13.6	12.8
60/80 mm	(%)	26.6	22.9
40/60 mm	(%)	43.5	42.7
30/40 mm	(%)	13.9	18.2
30 mm	(%)	2.5	3.4
Average diameter	(mm)	58.5	56.5
Bulk density	(kg/m^3)	445	456
Mechanical strength			
I 20		80.6	78.5
I 10		17.1	18.0
M 40		82.6	76.1
M 10		6.0	7.3
Average thickness of walls*	(m)	127	126
Average size of pores*	(m)	147	144
Porosity*	(%)	53.7	53.4
Reactivity (Japanese test I)*	(%)	29.00	29.05
Mechanical strength after reaction*	(%)	59.1	62.0

*Characteristics determined by BSC and BCRA

TABLE III

OPERATING RESULTS OF BLAST FURNACE No. 5

		Reference Period	Test Period
Type of coke			
Conventional coke	(%)	100	0
Test coke	(%)	0	100
Blast furnace operating characteristics			
Blast rate	(m^3N/h)	99,620	99,890
Blast temperature	(°C)	1,081	1,063
Coke rate	(kg/tci)	522.7	506.9
Fuel oil	(kg/tci)	48.6	46.7
Cast iron production	(t/day)	1,686	1,722
Blast Furnace availability	(%)	97.3	97.7

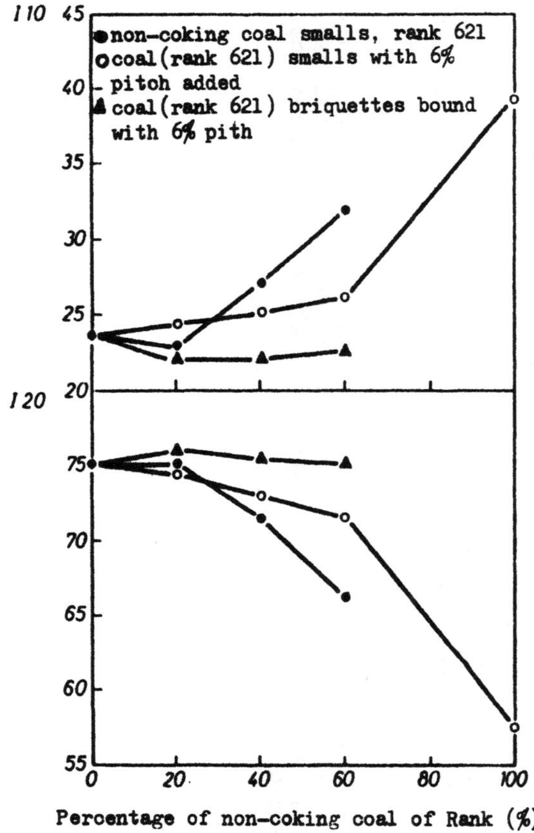

Fig. 1 : Trends in IR ID strength indices in terms of the percentage of non-coking coal introduced into the basic blend.

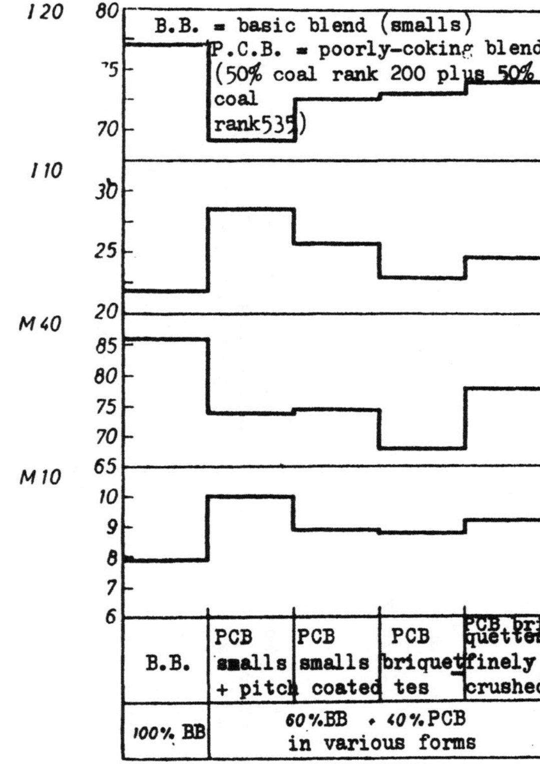

Fig. 2 : Effect of the form of introduction of the poorly-coking coal on the coke's mechanical strength indices.

FORMED COKE FROM HIGH-VOLATILE COAL VIA CHAR - THE CDF PROCESS

R. Guerin - Charbonnages de France
R. Degrave - H.B.N.P.C.
J. Verzeaux - H.B.L.

Summary

After 15 years of research one can state that technically valid processes now exist for producing formed coke both from low-rank coals as well as from high-volatile coals.

Because of the general economic situation none of these processes has been developed to date, though the constant increase in fuel prices is improving the competitive capacity of formed coke based on high-volatile coal, the char process becoming particularly more promising.

Formed coke from high-volatile coal via char - the CdF process

For 20 years research has been under way in numerous countries in the field of formed-coke production. In view of problems in the coal and steel markets since 1974 no industrial plant has yet been built. Research is still being conducted, however, and the Japanese have even organised a group of four major steel-producing companies to manufacture formed coke from high-volatile coal.

For 15 years Charbonnages de France, together with aid from the ECSC, have persevered with research in the field of formed coke and have obtained successful results from the technical point of view.

This report sets out to present the motivation for this research, the technical results and the economic calculations showing the economic conditions under which this process can be developed.

I. Research Motivation

For the northern part of the French coalfields, i.e. the HBNPC, the main motivation was a better way of upgrading their low-rank coal. Unfortunately, the very low price of fuel and coke prior to 1970 resulted in a rapid fall in low-rank coal production.

In 1966 requests were received from countries which had deposits of low-rank coal and consequently the HBNPC decided to commence laboratory investigation.

In 1969 encouraging results from the laboratory tests led to the decision to construct a pilot oven (1 x 0.60 x 6 m) at Vendin capable of treating approximately 10 t/d of coal.

In 1971 the HBNPC decided to construct a prototype plant with 4 chambers of 3 x 0.60 x 6.50 m capable of treating 150 t/d of coal. With this plant it was possible to produce enough coke to undertake a meaningful test into behaviour under blast-furnace conditions. In addition, this plant was capable of being copied as a full industrial installation. This prototype was capable of producing a satisfactory type of formed coke, not only from low-volatile coal (200) but also from high-rank bituminous coals (711).

In 1969 the HBL constructed a rotary pilot oven for the production of a special type of coke for the ferro-alloy industry, the purpose being to upgrade Lorraine coal. In 1972 an industrial-scale rotary oven went into service at Marienau with a capacity for treating 9 t/h of Lorraine coal and in 1979 a second oven went into production at Carling with a capacity of 24 t/h; both ovens produce electrometallurgical coke.

Tests on the production of char based on Lorraine coal, which were carried out in a prototype plant with a capacity of 300 kg/h at the Marienau station, were confirmed in the 9 t/h oven. The char could be transported in standard rail cars to Drocourt for the production of 350 t of pellets which, after carbonisation in the prototype plant, produced a highly satisfactory quality of formed coke.

These first steps taken by CdF correspond with the three types of motivation which had appeared through the years in the field of coking, namely:

a) to produce a type of coke suitable for blast furnaces

and in fact to aim at a performance which is superior to that of conventional coke by improving the capacity to cope with production parameters: predetermined shape, size and structure of the pellets, regularity of a continuous production process.

b) to use coal which does not fuse

The conventional coke oven is seeing an ever increasing range of coal capable of being used in large quantities as charging blends due to the introduction of processes such as preheating, briquetting and stamp-charging. For coke, however, 'transitory fusion' must be applied in the oven to allow for the porous structure of the coke.

Formed coke, on the other hand, requires the use of coal which does not become deformed during the coking process in order to prevent clustering of the pellets. It so happens that the world possesses enormous reserves of high-volatile coal which is not liable to fusion, and this does not even include the lignite range. It would, therefore, be beneficial for the future to develop a method for producing a satisfactory quality of coke based on these raw materials which will be available at relatively low cost for a considerable period of time.

c) to produce coke at less cost and under better working conditions

It is not necessary to spend much time in a coking plant before one becomes aware that despite major steps taken over the last 10 years to reduce pollution the working conditions remain arduous for workers in charge of the coke-oven machines (working rates and high temperatures) and for those responsible for maintaining the temperatures of the refractory surfaces. With formed coke one can operate as with a fully-enclosed chemical reactor, it being only necessary to handle cold products.

In addition, with the HBNPC formed-coke oven one can shut down and start up operations as desired, a facility which makes it possible to adapt to economic fluctuations and consequently to reduce the average annual price of the coke produced.

In short, one should not condemn formed coke too soon, since it may turn out to have a rosy future.

2. Technical Results of CdF Research

The CdF have been conducting research into formed coke with ECSC aid since 1971.

2.1. Direct carbonisation of high-volatile coal

Research into the direct carbonisation of high-volatile coal has been carried out in the laboratory, in the pilot oven and in the Drocourt prototype oven. We shall not attempt here to give details of this work but instead shall try to present the directions this research has taken over the years.

Laboratory

a. Tests were conducted with 40 gramme pellets briquetted to 250 Kg/cm^2 with 10% coal pitch from high-rank blends (711) VM_{dry} = 36.5 and SI = 1½ and medium-volatile coal (433) with VM_{dry} = 24 and SI = 8½.

The following results were obtained:

Blend	Strength (°/mn)			ASPECT		
	10	7	3	10°/mn	7°/mn	3°/mn
100 0 10	58	30	22	Deformed	Good	Good
90 10 10	38	32	28	Slight deformation	Slight deformation	Good
80 20 10	81	65	44	Oblique deformation	Oblique deformation	Good
75 25 10	81	100	70	Deformation	Slight deformation	Good
70 30 10	67	67	91	Cracked. Deformed	Oblique deformation	Good
60 40 10	46	68	>100	Cracked. Deformed. Swollen	Deformed. Swollen	Good

b. Tests were conducted with 1.6 Kg of pellets (40 g) in a special oven with different percentages of medium-volatile coal and at different carbonisation rates.

These two series of tests show that with the two parameters selected, the heating rate and the percentage of medium-volatile coal, there is a wide variation in

DIAGRAM No. 1

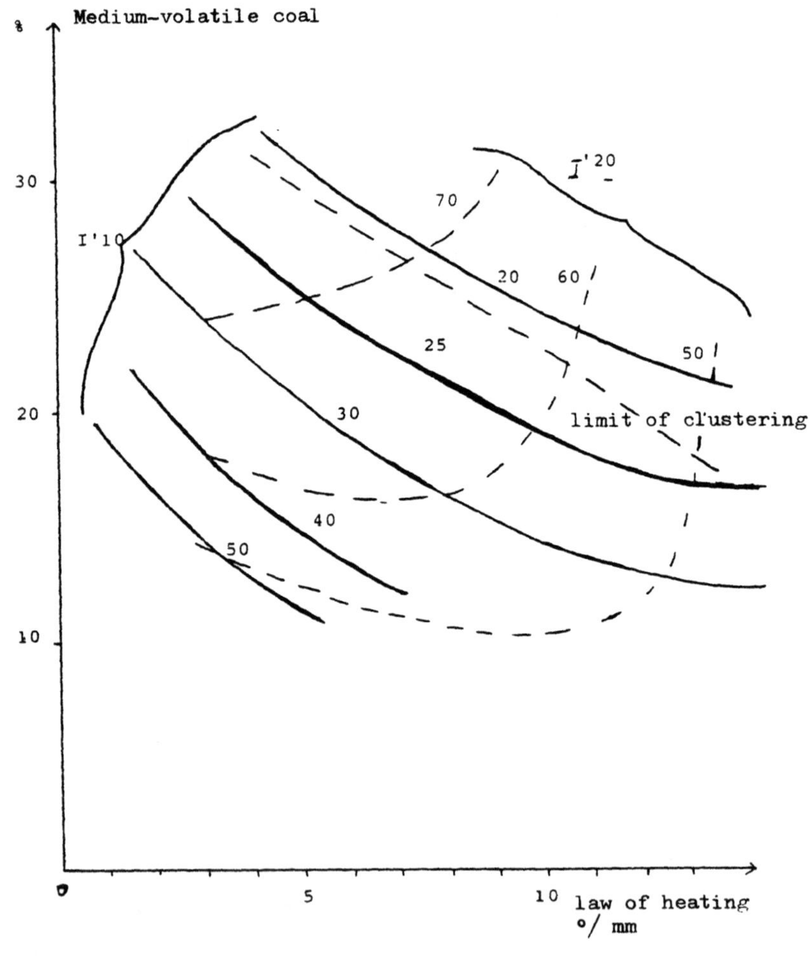

the quality of the coke produced, a fact which assists in determining the research to be carried out in the pilot and in the prototype oven.

Three features can be established:

- When the proportion of medium-volatile coal is increased the pellets quickly become deformed, this resulting in congealing. The maximum proportion of medium-volatile coal which may be introduced decreases rapidly as the coking rate is increased.
- Coke cohesion represented by I'10, the index derived from I_{10}, improves as one increases the proportion of medium-volatile coal and the coking rate.
- An attempt has been made to determine coke fragmentation. This is associated with pellet fissuring, their crushing strength with the proportion + 20 mm and measured by I'20, the index derived from IRSID. The latter increases with the proportion of medium-volatile coal and with a slower heating rate.

The results are shown at Diagram No. 1.

For producing formed coke one, therefore, requires a heating rate which is sufficient for good coke cohesion, however one is soon limited by fissuring. It then becomes necessary to increase the proportion of medium-volatile coal, though here one is restricted by the tendency to cluster which impairs the operation of the oven.

The production of formed coke, therefore, represents a compromise which is dependent on the coal used and on the configuration of the plant.

<u>Pilot plant</u>

75 tonnes of 30 g pellets were passed through the pilot plant at the rate of 10 t/d with two blends, one having 88% low-volatile coal (200) and 12% medium-volatile coal (433) and the other 88% high rank coal (711) and 12% medium-volatile coal (433); different coking rates were applied.

The following results were obtained:

Low-volatile	88			
Medium-volatile	12	12	12	12
High-volatile		88	88	88
Rate ^{o}/mn	7	7	5	4
Coke - 25 mm	11	31	24	17
Porosity	32	42	42	42

These results confirm the principle that in order to obtain a satisfactory fragmentation strength a lower coking rate must be applied.

Prototype plant

With the information obtained from the pilot plant it was possible to move to the prototype plant and to select coking rates, proportions of medium-volatile coal and pellet size (37 g) in such a way as to eliminate production problems and obtain an optimum coke quality.

Low-volatile	88			
Medium-volatile		12	12	22
Rate ^{o}/mn	10	6	5	5
VM_{dry}	4.5			2.2
I 10	21-25	38	42	22-25
- 27 mm	15	24	19	18
Porosity	33			45

The results are more favourable than in Diagram No. 1 because there is a lesser degree of variation due to the walls.

It is, therefore, possible to find a compromise allowing the production of a satisfactory quality of coke by passing it directly through a Drocourt-type oven, though the coking rate is low, a fact which will mean higher production costs particularly on account of the size of investment required.

2.2 Char-based carbonisation

The end to the production of high-volatile bituminous coal at the HBL resulted in the HBNPC conducting research based on char produced from high-volatile Lorraine coal.

Laboratory tests showed that viable formed coke can be produced at a coking rate of 10°/mn.

As with formed coke produced directly from high-volatile bituminous coal the influence of the law of heating at higher coking rates and the proportion of medium-volatile coal added to the char is large (cf. Diagram No. 2).

As is the case for formed coke based on char one sees that it becomes necessary to add about 20% medium-volatile coal (SI = 8), that the char does not 'wet' well and that the proportion of pitch must be increased to 12% for briquetting.

In 1978 the HBNPC produced 350 t of 50 g pellets from char, these pellets lending themselves well to a coking rate of 10°/mn.

The coal used for making the char was Lorraine high-volatile coal (632) and medium-volatile coal (433) with the following characteristics:

	VM_{dry}	SI
632	35	3
433	20	8
char	10	1

In this test formed coke with 88% low-volatile coal and 12% medium-volatile coal was compared with char-based formed coke. The following results were obtained:

	Coke from char	Coke from LV coal
I 10	27.6	25 - 27
% - 25 mm	24	15 - 20
Porosity	43	32
V M	1.8	1.9

One could improve the - 25 mm to less than 20% by reducing the size of the pellets to 40 g instead of 50 g and by adding tar to the briquette.

The value obtained for I 10 may appear high to those using conventional coke screened to 40 mm or even to 60 mm, however the tests were carried out with conventional blast-furnace coke (+ 27 mm) and produced an I 10 value of 27.

The CdF group, therefore, succeeded in producing formed coke with satisfactory properties using high-volatile coal via char.

2.3 Blast furnace tests

Output from the Drocourt prototype plant was used for the blast-furnace tests.

The main results obtained from these tests are given below, together with the most important aspects of the operation.

Year	Blast furnace	Diameter	% Formed coke Average	Max.	Formed coke T	Duration of test	Conventional coke (Kg/t cast iron)	Formed coke (Kg/t cast iron)
1973	Longwy	5.6	25.5	32	3,280 (1)	30 days	582	593
1973	Denain	7	27	43	5,675 (1)	31 days	540	540
1975	Dunkerque	9.5	25	44	11,145 (1)	29 days	492	486
1976	Dunkerque	9.5	26.5	27	26,310 (1)	58 days	495	512
1978	Denain	9.5	29.4	31.5	19,850 (2)	40 days	487	487
1978	Decazeville	4	38	58.4	2,440 (1)	44 days	528	522

(1) Pellet weight - 40 g (raw) 32 g (carbonised), 27 x 37 x 47 mm
(2) Pellet weight - 50 g (raw) 40 g (carbonised), 32 x 43 x 51 mm

During the tests the blast furnaces behaved in a manner which was comparable to that when conventional coke is used, this information having been supplied by blast-furnace operators taking part in the tests.

Numerous tests were conducted to compare the quality of formed coke with conventional coke. The characteristics being compared are shown in a Table which makes use of data from an article by M. NISHIDA (Fuel Society of Japan, 1.11.76) on processes which differ from those used by the CdF:

	Hot briquetting	CdF LV coal	CdF H V direct	CdF H V via char	Cold briquetting	Conventional coke
Porosity	28	32	45	45	44	50-60
VM dry	10.5	1.9	2.2	1.8	3.9	0.3-1
DI^{150}_{15}	56.7	76-82	76-80	74	86.9	80-85
Strength after reaction	35.8		39		8.1	50-70
T reaction threshold	800			760		920

From the Table one can see that the properties of the formed coke obtained by the CdF group from low-volatile or high-volatile coal are not very different (with the exception of porosity) and that they can be compared with the properties of conventional coke; from this one may suppose that the characteristics of formed coke produced from high-volatile coal will be suitable for blast-furnace use in the same way as formed coke produced from low-volatile coal.

3. Economic Study

One way of making an economic comparison of the processes is to calculate the price of one tonne of blast-furnace coke while making assumptions on the level of investment, the value and output of the gas, tar and small coke and how this affects production costs and the price of the raw materials: pitch and low-cost coal.

	wet gravity	wet stamp charging	formed coke		
			from LV coal	from HV coal direct	via char
Raw materials for 1 T BF coke					
333	269	286	143		
433				340	329
434	672	286			
632	403	868			
200			1,037		
pitch			110	200	161
711				1,220	1,256
VM dry	27	30	15.5	35.7	20.7
Running costs	73.7	80	106	137	152
Depreciation	202.3	202.3	153	293	220
Returns	167.4	175.8	161	349.5	359

The returns are calculated for a value per therm (net calorific value) of gas of 0.1 F/Th for processes in conventional ovens and 0.08 F/Th for processes using formed coke.

From these values one can draw up the graph shown at Diagram No. 3 which compares the price of blast-furnace coke produced by the different processes as a function of low-cost coal.

Diagram No. 4 presents a parallel comparison by taking the value per therm to be 0.2 F/Th and 0.16 F/Th respectively.

An examination of graph No. 3 (conventional coke oven gas at 0.1 F/Th) shows that:

a) it is confirmed that wet stamp charging produces a coke which is less expensive than gravity charging if the coal is cheap and of low quality;
b) formed coke produced directly from high-volatile coal is not competitive.

Formed coke produced from high-volatile coal via char is competitive if a low-cost coal is used with the wet gravity method. It is not competitive with stamp charging if the same type of coal is used, though it could become so in a country which produces coal suitable for formed coke and which would have to import coal for stamp charging, represented for example by the point x = 300 F/t.

c) on examination of graph No. 4 which represents a very high fuel price (0.2 F/Th for conventional coke oven gas) one sees that the production of formed coke via char is the least expensive and that the direct production of formed coke is only competitive for a mining country where coal has to be imported for conventional ovens;

d) formed coke produced from low-volatile coal is more competitive at low fuel prices.

DIAGRAM No. 3

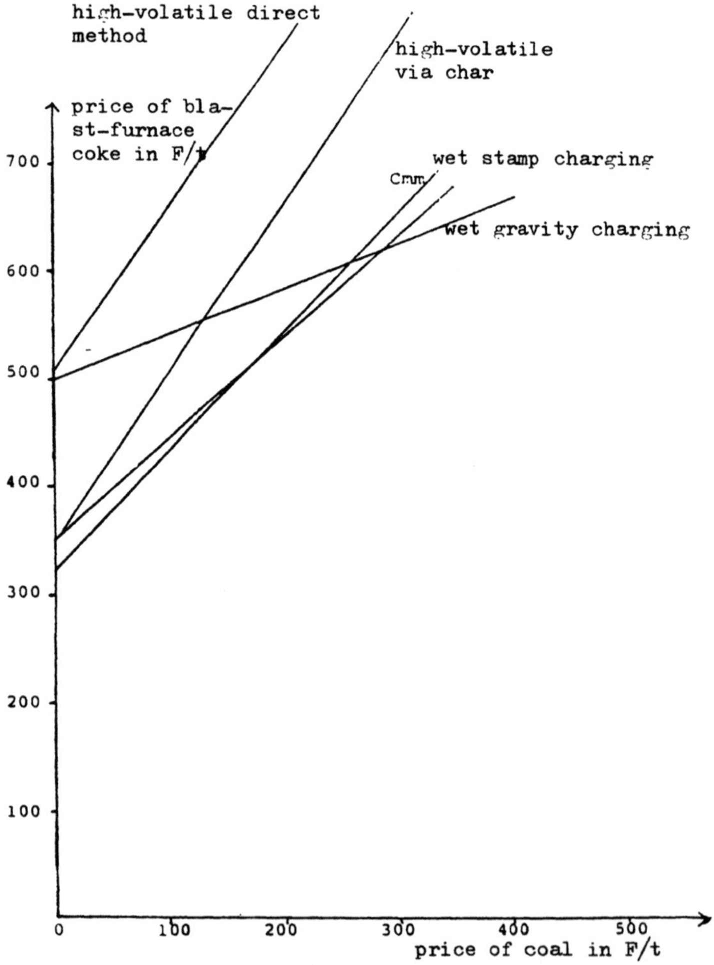

DIAGRAM No. 4

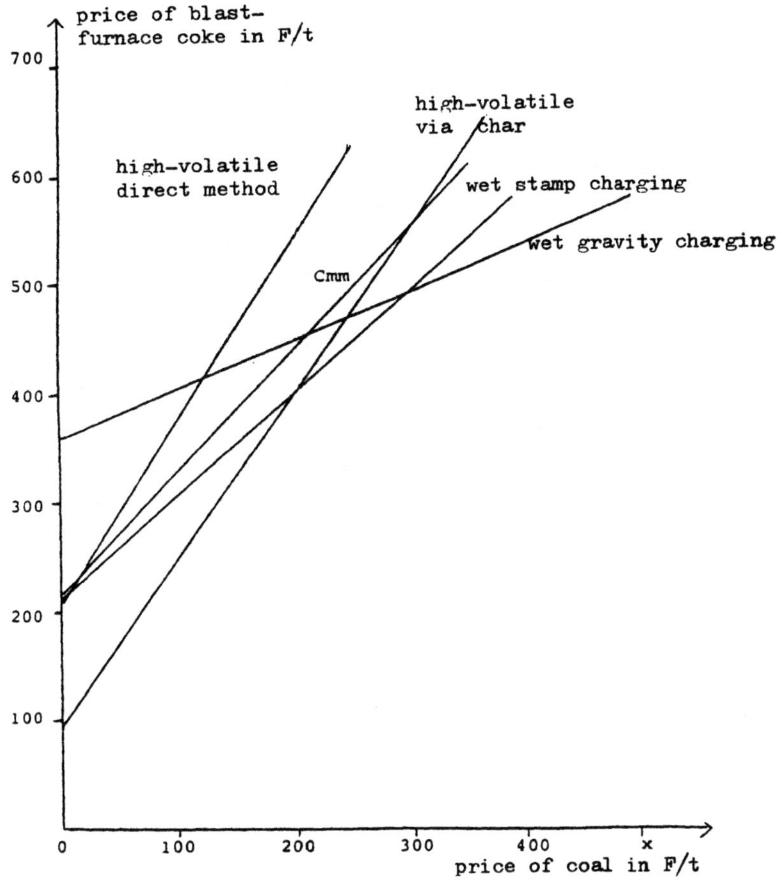
price of blast-
furnace coke in F/t
700
600
500
400
300
200
100
0
100
200
300
400
x
price of coal in F/t
high-volatile
via char
high-volatile
direct method
wet stamp charging
wet gravity charging
Cmm

Conclusion

The research work conducted by CdF over the past 15 years has led to the development of technically-satisfactory processes for producing formed coke; these processes have been tried out on an industrial scale and have proved capable of producing a satisfactory blast-furnace coke.

In economic terms the processes for producing formed coke cannot develop under present conditions in the coal and fuel markets. As regards formed coke produced from low-volatile coal those mining countries which have LV coal usually possess other types of coking coal and in most cases pithead prices do not favour low-volatile coal. What is more, low-volatile coal can be put to better use in domestic grates.

As for the other processes based on high-volatile coal, only that to produce coke via char seems to be competitive, and this process may prove useful for very cheap types of coal or lignite which are not suitable for conventional coke ovens.

The assumption that the price of fuel will increase even more sharply in the years to come tends to favour the production of formed coke from high-volatile coal via char - this becoming the most economic process. Consequently, longterm research should be aimed in this direction.

THIRD TECHNICAL SESSION: COKING TECHNIQUES AND SUMMARY OF ROUND TABLE DISCUSSIONS

Chairman: K.G. BECK, Mitglied der Geschäftsführung der Bergbau-Forschung GmbH

- Thermal measurements on coke ovens
  D. SARANGI, Bergbau-Forschung GmbH

- Control of coke oven heating
  J.P. GAILLET, Centre d'Etudes et Recherches des Charbonnages de France

- Testing of the Codeco system for the programmed heating of coke ovens
  D. HABERMEHL, Bergbau-Forschung GmbH

THIRD TECHNICAL SESSION

Chairman: K.G. BECK

SUMMARY OF THE DISCUSSION

Mr MONSON noted that the coke guide would obviously have to be moved out of the way so that the camera car could be positioned in front of the open coke oven. He asked whether this did not give rise to operational difficulties.

Mr SARANGI confirmed that it was for this reason also that preference had been given to measurement on the coke side. Careful preparation was important, and on that basis measurement could be carried out in two minutes. Thus in practice no operational delays occured.

Mr FLASCHE asked whether the measurement results were not sometimes distorted by the fact that graphites deposits burnt if the oven remained open for the purpose of measurement.

Mr SARANGI said that in his experience such effects could indeed occur, but should then be clearly brought out on the thermogram, and were usually found on the upper parts of the walls, where temperature rises of that kind were generally not to be regarded as anomalies of oven heating.

Mr BECK asked exactly where the infra-red pyrometers were installed, and whether they were adequately protected against heat and dust.

Mr CAILLET replied that the pyrometers were mounted behind the sides of the coke guide, so that the surface temperature of the expelled coke cake could be measured through small openings in the sides. The pyrometers were placed about 1 m away from the coke cake surface and were also cooled using a fan.

On the report by Mr Habermehl, Mr GAILLET first asked whether the reduction of the heat consumption tallied with the lower final temperature of the coke registered with programmed heating. Mr HABERMEHL reminded the meeting of the slide showing the Sankey diagrams for conventional and programmed heating. With programmed heating, the final temperature of the coke was reduced by about 70 K and the oven free space temperature by about 20 K; i.e. the sensible heat of the produced during carbonization was also reduced. From these two figures one then calculated the reduced heat loss, which corresponded to the heat consumption ascertained.

Mr ECHTERHOFF asked whether, for the periods with raised heat input using 170% or 110% of the normal gas quantity, the values were empirically ascertained or obtained from a mathematical model. In addition, Mr ECHTERHOFF wanted to know whether programmed heating had further development potential, going beyond the results hitherto announced in terms of reduction of energy input. Mr HABERMEHL took the view that the possible effect had been largely achieved, since the considerable saving resulted from a reduction in the sensible heat of the coke and the crude gas, which could not be reduced at will. The percentage saving would of course decrease in inverse proportion to the care with which the undergrate firing system was adjusted and monitored with conventional heating methods. The result from the experimental coking plant had shown this. The heating programme was indeed derived from a pyrometric model, but the block heating system presented here and found to be suitable was the result of the experiments and development work carried out at the experimental coking plant.

Mr de Torcy put three further questions to Mr Habermehl, which were answered as follows:

1. What was the criterion for coking time?

   Since no generally valid criterion was yet available at the experimental coking plant, the coke cake was pushed 'at the 11th hour', i.e. the coking time was so short that small portions of green cokes still occurred in the coke cake. As a check, the ejection of the coke cake was recorded with a television camera to obtain as objective as assessment as possible.

2. How much scavengingair was used during the heating pauses, e.g. as a percentage of t he air quantity fed into the oven in conventional heating?

The quantity of scavenging air was kept as low as possible, since it led to further heat losses which prejudiced the desired result. Depending on the heating system wed, 5 to lo % of the air quantity needed with conventional heating was introduced as scavenging air.

3. What was the heating pattern for an oven which was at the junction of two oven blocks, in programmed heating, run together on one programme?

Extensive studies had been carried out on this, since with the three experimental ovens and their equipment this question could not be experimentally investigated. However, it was planned to carry out this experiment in the near future on a block of ten ovens. But the result of the studies already carried out showed that no fundamental difficulties were involved, since there was only one hour's difference between the two blocks at the end of the coking process. For the so-called "boundary oven" this meant that the heating of one wall of the oven took place an hour earlier or later than the other in the heating programme, but that this oven as a whole retained the same, adequate, amount of heat as the others. Given the known pyrometric inertia of the coke oven, no difficulties were anticipated in this connection.

THERMAL MEASUREMENTS ON COKE OVENS

B. Sarangi

Bergbau-Forschung GmbH

Summary

Current methods of temperature measurement on a coke oven are set out and the inevitable inaccuracies and limited information content of these measurements assessed as regards the thermal monitoring of coke ovens. It seemed desirable that a new measuring method be found which would determine the temperature distribution on the chamber walls of horizontal chamber ovens using infrared thermography.

After extensive investigations, a measuring device was designed and constructed which allows swift and precise positioning of the infrared camera in front of the door opening on the coke side and recording and processing of the measurement data (1).

Allowing for chamber geometry and photographic conditions (positioning and travelling speed of the infrared camera), a computer programme was developed to produce the thermograms.

Working from the techniques available at present and the knowledge acquired in the course of the above work, a more advanced infrared measuring device was built with the help of a specialist firm. This new measuring technique evolved by Bergbau-Forschung GmbH, Essen, can be of valuable assistance in providing a modern monitoring system for coke ovens and in carrying out thermal examinations on coke ovens.

Introduction

Coupled with the further development of coking techniques, great importance is being attached to monitoring systems, and particularly in this case to thermal monitoring of coke ovens.

Temperature monitoring at the coke oven is carried out generally by measuring the flue temperature at a set time with an optical pyrometer. However, these measurements only record the local temperature of the nozzle brick in certain waste gas flues and this provides no information as to the mean or maximum temperature of the chamber wall. In addition, the geometrical location of the nozzle brick differs according to the type of heating system, so that nozzle brick temperatures of the various oven systems cannot be compared with one another despite comparable operating conditions.

To record the vertical temperature distribution in the flue, there are several methods (3,4) which involve introducing suitable probes. Because of the high cost of equipment, however, these methods have not become a part of operating practice. As an additional form of temperature recording, chamber lining measurements (5) are taken in order to register the temperature/time curves in the coke oven. The vertical temperature profiles at the chamber centre towards the end of the coking time, as shown in Figure 1, are used here to assess the setting of the heating system. Irregularities in the lining or non-vertical insertion of the measuring probes, however, may result in an invalid set of measurements.

As a supplement to these measurements, after pushing of the coke the chamber wall temperature is measured with the doors closed through the oven charge holes at two points, namely 400 mm above the oven floor and 400 mm below the top, using an optical pyrometer. In practice, however, measurements are taken at only three charge holes to save time, so that the mean chamber wall calculated from six individual values represents only a very approximate value in view of the large chamber wall area. For this reason, other methods have been developed to increase the density of measuring points and to allow continuous measurement of the chamber wall temperature (6 to 8). Finally, we should also mention a measuring process used in France in which in contrast to the measurement of the chamber wall temperatures the thermal radiation of the coke cake during the pushing process is recorded with the aid of one or two infrared pyrometers. The temperature profiles measured are used to control heating. But all these methods have a disadvantage in that the measurements are made at only one particular level; they can only provide information on the temperatures in the horizontal direction from the machine side to the coke side and not in the vertical direction from the oven floor to the gas collecting chamber.

Problems and objectives

In view of the inevitable inaccuracy and limited information content of the traditional methods of measuring the temperatures of the coke oven walls, it seemed desirable to develop a more suitable measuring technique. A good candidate was infrared thermography which for several years has been the state-of-the-art method of recording temperature distribution over large areas (9,10).

The particular problem in measuring the temperature distribution on the chamber walls of coke ovens lies in the fact that the areas to be measured are relatively inaccessible. The photograph must be taken from the outside through the door opening and the geometry of the chamber is such that the optical conditions for the photograph are poor. This has considerable consequences both for temperature measurement itself and for assigning the correct coordinates to the measured values.

Measuring procedure

The term "thermography" is used to describe the graphic representation of temperature distributions over large surfaces. Infrared thermography is based on the quantification of infrared radiation which is emitted from every object and which, in accordance with radiation laws, can be equated to a particular temperature (11).

The infrared cameras commonly used to record temperature fields operate on an optical/mechanical scanning system which consists of oscillating and rotating mirrors or prisms and which divides the area to be photographed into chronologically ordered picture elements. These picture elements are continuously fed to a photo-electric radiation-responsive receiver (detector) and according to the intensity of radiation of each, they are converted into degrees of light, with the result that a grid-shaped black and white or even colour picture of the temperature distribution appears on the screen.

Figure 2a is a schematic diagram of the structure of the infrared camera (thermoscan) used in the tests. The radiation from the object reaches a tilting plane mirror /1/ which scans the object in a vertical direction. The radiation is then reflected onto a rotating mirror /2/ which scans it horizontally, and finally reaches a parabolic mirror /3/ which focuses it on the detector /4/. The detector transforms the in-coming radiation energy into electrical energy which, after the appropriate pre-amplification /5/, is passed on to a picture reproduction unit. A reference radiator /6/ is built into the camera unit to determine the absolute temperature.

Due to the unavoidably small gap between camera and door-frame, an aperture angle $w \geq 120^{o}$ is absolutely essential for scanning the entire chamber wall. By modifying the measuring head, the infrared camera used here can also be employed as a line scanner (with an aperture angle $w = 120^{o}$). Figure 2b shows the camera as a line scanner.

It can be seen from Figure 3 that observation of the entire chamber wall area is only possible through the door opening (machine or coke side). For this reason, the measuring equipment is positioned in such a way that it is possible to take a photograph of the entire chamber wall through the door. The infrared camera is positioned at a level equal to half the chamber height, so that

- the direction of viewing coincides with the chamber diagonal

(Figure 4), and
- the full height of the chamber on the coke side can be covered at a constant aperture angle of 120°.

The infrared camera is situated on a movable stand and is driven past the opened chamber at a prescribed speed, while the temperatures in the direction of the chamber verticals are constantly measured. The measurement data are recorded on a magnetic tape.

Figure 5 is a schematic diagram of the measuring device; it consists of:

- an infrared camera with a monitor and the associated electronics,
- control and recording equipment, and
- a lower section which can be moved on rails and on which is mounted a camera sled adjustable in three directions. In this way, swift and precise positioning of the infrared camera can be guaranteed.

Essentially, taking one measurement involves three operations:

- Adjustment of the infrared camera to the measuring position dictated by the oven geometry,
- Precise positioning of the camera sled in front of the chamber to be measured, and
- Scanning of both the wall surfaces to be measured.

To adjust the infrared camera to the measuring position, it is necessary for the parallels to the longitudinal axis of the group of ovens and to the longitudinal axis of the oven chamber to be near the measuring trolley or outside the oven group. Once this measuring work is complete, the camera sled can be positioned. Then the measuring proper begins.

The previous measurements were carried out with attention to the following aspects:

- Proof of reproducibility. To this end, the chamber walls of an oven were in each case scanned twice in quick succession from the same position. In between the various measurements, the camera is merely moved to and fro in the horizontal direction on the sled and the angle transducer actuated.

- The effect of fine adjustment. In view of the fact that this measuring technique was used during actual operation, the two chamber walls of the same oven were measured after successive coking periods.
- The effect of cooling down. The same chamber wall areas were scanned at various times (5, 7.5 and 10 minutes) after the first measurement with the door on the coke side open.

Analysis of the measurement data

The analog measurement data together with trigger signals are recorded on the magnetic tape and describe the area shown in Figure 6. Within this area, the portion of chamber wall, i.e. the useful portion of the camera signal, varies according to the location of the camera, in the example given it increases from around 14% on the machine side to 100% on the coke side. When the chamber wall is scanned vertically, a strip is scanned the location of which will depend on the current field of vision of the camera; the width will also depend on the position of the camera with respect to the chamber wall.

The aim of this evaluation is to produce a thermogram of the chamber wall. Thus, the first step was to establish a grid for dividing up the chamber wall. Due to the height/length ratio of approximately 1:3 for the chamber wall and the attainable resolution, a grid of 120 rows and 40 columns was selected, i.e. the chamber wall is covered by 4800 grid elements. Correction of any distortion occurs in two stages:

- Correction of horizontal distortion. 120 scanning strips are taken from the measurement data by suitable grouping.
- Correction of vertical distortion. As shown in Figure 6, for each of the 120 strips that portion of the measurement data is removed which is allocated to the chamber wall (useful section).

Considerable mathematical effort is necessary for this transformation of the perspective distortion of the chamber wall to the grid, so that evaluation of the measurements is only possible using a computer.

The thermogram of the chamber wall is calculated from the measurement data with the aid of a process computer and is output on a printer or, in colour, on a graphics terminal (Figure 7).

Discussion of the measurement results

Reproducibility: 86% of the absolute differences in temperature between the individual picture elements of a chamber wall (differential thermogram) which were measured twice from the same position in quick succession were within the temperature resolution limits of the infrared camera. Thus, good reproducibility is guaranteed with rapid repeat measurements.

Fine adjustment: Similarly, the differential thermogram of the same chamber wall was produced here after successive coking periods (under comparable operating conditions as regards charge coal, charge weight and heating). In this case, 91% of the temperature differences between the various measurements were within the resolution limits. This confirms the suitability of the measuring process for general practice.

Cooling: An expanse of colder wall areas starting from the coke side can be read from the thermograms of the chamber wall which were taken when the door on the coke side was open. The hot wall sections cool down simultaneously.

Analysis of the cooling-down of selected sections of chamber wall has shown that it is possible within the first ten minutes to compute the initial temperature distribution on the chamber wall (at the time of pushing).

Conclusions and prospects

The method described not only allows the actual temperature directly on the chamber walls to be determined but also enables the temperature distribution over the entire chamber wall area to be recorded as it were "at a glance". It is thus now possible to quickly detect any temperature anomalies that may occur during operation in certain areas of the chamber wall and eliminate such anomalies by appropriate action.

Furthermore, this method, developed by Bergbau-Forschung GmbH, Essen, permits the temperature radiation of the entire coke cake surface during pushing to be photographed through a small slit in the coke guide.

The method also enables surface temperatures, for example of an entire oven group (roofs, doors, buckstays) to be determined; such surface temperatures are essential for quantifying the surface losses and for drawing up a thermal balance sheet for the oven group.

The method can also be used for general monitoring purposes - locating areas of over-heating, lack of insulation, leaks etc.

At this point, we should like to thank the Executive Board for Energy of the Commission of the European Community for their financial assistance in this work.

LITERATURE

1) B. SARANGI:
Development of a method for measuring and recording temperature distribution on the chamber walls of horizontal chamber ovens.
Dissertation, Technical University of Clausthal, 1980.

2) W. EISENHUT and R. WORBERG:
Heat consumption and heat control in coke production.
Ironmaking proceedings 38 (1979) p. 347-53.

3) E. PRIES and F. WACKERBARTH:
A new measuring method.
Ironmaking proceedings 35 (1976) p. 510-12.

4) U.M. SCHEDL:
Heat and temperature distribution in a coke oven flue.
Dissertation, Technical University of Berlin, 1972.

5) G. JURANEK, A. ADELSBERGER and W. TODROWSKI:
Temperature measurements in coke ovens with probe bundles.
Glückauf-Forschungshefte 30 (1969) p. 79-83.

6) W. WESKAMP and E. SCHIERHOLZ:
A device for measuring the chamber wall temperature of coke ovens during pushing.
Glückauf 105 (1969) p. 6-8.

7) L. BUGLAK:
Instrument for measuring oven wall temperatures.
Coke and Chemistry 1969 (7) p. 17-18.

8) M. ORLIK, T. LONGIN and J. MOLINEK:
Temperature measurements on heated walls of coke oven chambers.
(Czech.) Hutn. Listy 32 (1977) p. 235-240.

9) Proceedings of biennial infrared information exchange (IRIE '74, '76, '78).
AGA Corporation, N.J. 07094, USA.

10) Lectures on the occasion of the special infrared symposium.
German Society for non-destructive testing, Dortmund (1975).

11) GLOCKMANN, W.:
"Thermography" in the handbook of technical temperature measuring.
Publisher: F. Lieneweg, Friedr. Vieweg & Sohn Verlagsgesellschaft mbH, Braunschweig (1976), p. 436-444.

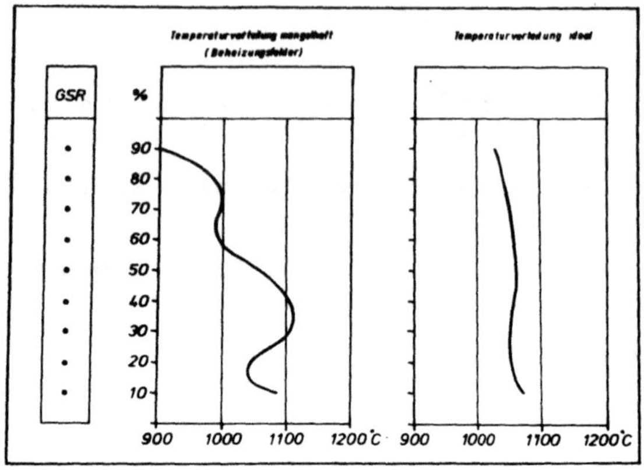

Figure 1. Temperature distribution in the chamber lining at the end of the coking period. Temperature distribution poor (heating fault); Temperature distribution ideal (right).

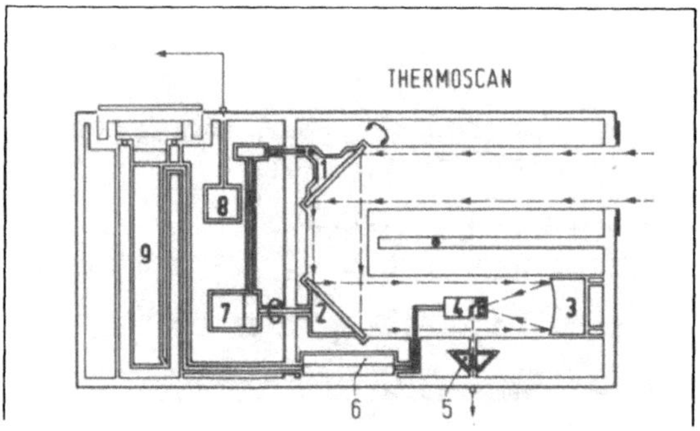

1 Tilting plane mirror
2 Rotating mirror
3 Parabolic mirror
4 Detector
5 Preamplifier
6 Reference radiator
7 Motor
8 Synchronizing signal
9 Nitrogen Dewar vessel

Figure 2a. Functional diagram of the infrared camera.

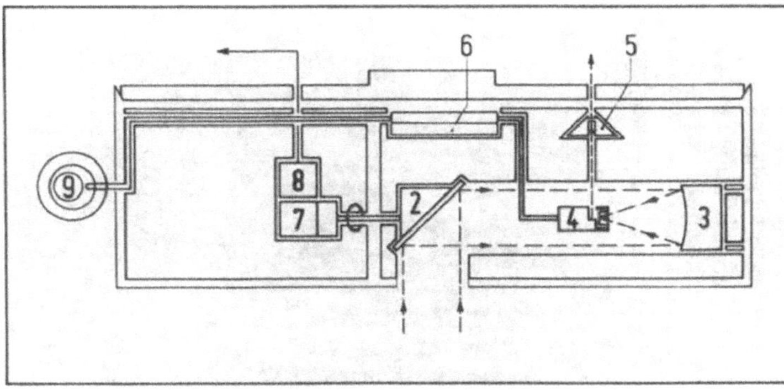

Figure 2b. Schematic diagram of the infrared camera as a line scanner.

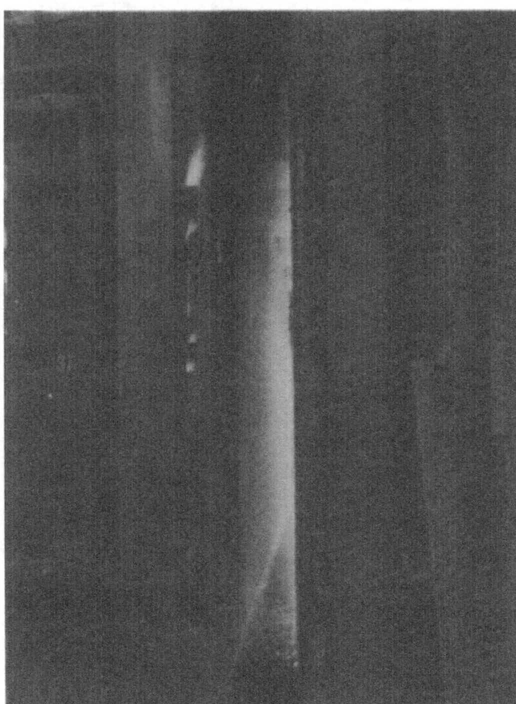

Figure 3. View into a horizontal chamber oven.

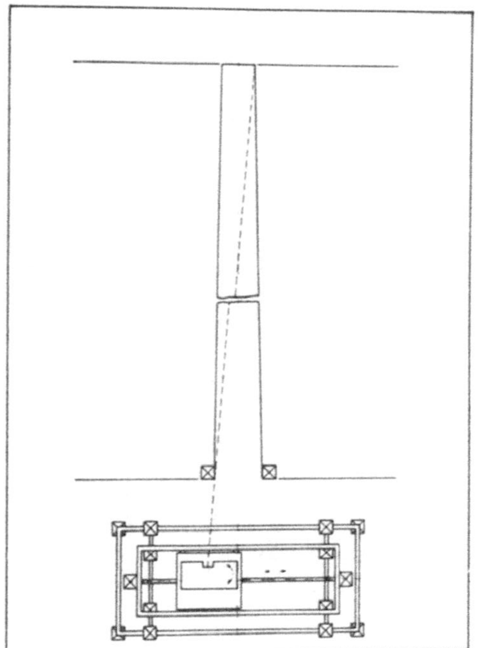

Figure 4. Geometry of the scanner photograph.

1 Infrared camera
2 Monitor
3 Magnetic tape device
4 Oscilloscope
5 Control unit
6 Power supply
7 Chamber wall

Figure 5. Setup for the chamber wall thermogram.

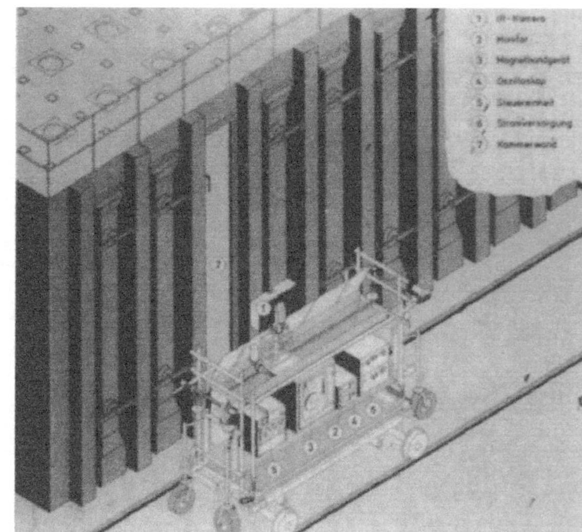

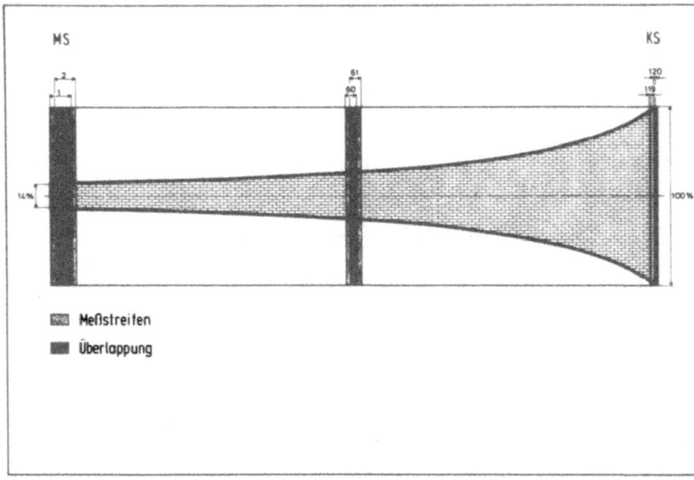

Figure 6. Perspective distortion of the chamber wall with respect to the entire measurement area. Measured strip; overlap (black).

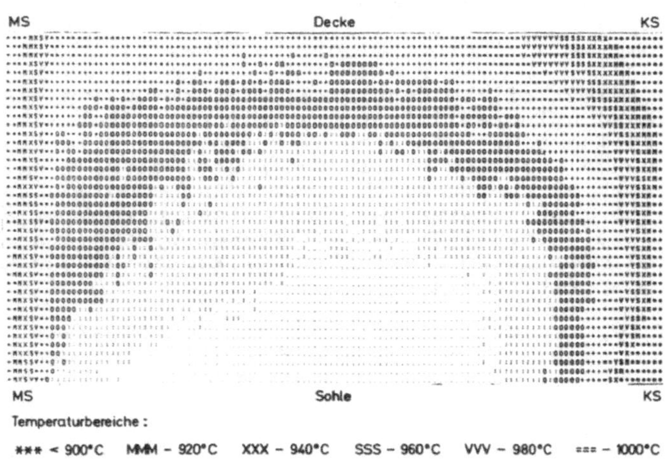

Figure 7. Thermogram of a coke oven chamber wall. Roof (top), floor MS; machine side KS; coke side; temperature ranges (bottom).

CONTROL OF COKE OVEN HEATING

J.P. Gaillet

Marienau Experimental Station

Summary

The Marienau Experimental Station has undertaken, with Community aid, a study aimed at improving monitoring of the thermal regime in coke ovens by, on one hand, devising new representative criteria and, on the other hand, developing new fixed measuring sensors which will provide continuous recordings centralised in a control room.

In this paper, the author describes three new methods of measurement, whose reliability and representative character have been amply demonstrated in various series of tests at coking plant.

Measurement of the temperature of the refractory brick by means of thermocouples inserted into the crown of the ovens provides a reliable assessment of the battery's temperature, and can replace manual readings obtained by means of an optical pyrometer at the bottom of the flue.

Measurement of the coke's temperature during discharging of the ovens, using an infrared pyrometer installed in a fixed position on the coke guide, enables the coke's surface temperature to be recorded easily and systematically along the entire length of the oven, thus providing complementary information to that obtained by the optical pyrometer as described above. Installation of infrared pyrometers at several levels also offers the possibility of assessing the quality of vertical temperature distribution.

Finally, measurement of the distribution gases' characteristics in a rising column enables the degree of carbonisation of the coke to be determined. Developments in the temperature and in the hydrocarbon (particularly methane) content are indicative of the degree of progress of coking, and can be correlated with the mean temperature of the cake or, even better, with the temperature of the lesser-carbonised zones.

I. INTRODUCTION

Temperature regulation of coke oven batteries is one of the decisive parameters in metallurgical coke production. It has a direct influence on the uniformity of the coke's mechanical properties, the thermal efficiency of the ovens, cell productivity and the life of the refractory linings.

Temperature regulation is generally monitored by measuring the temperature at the bottom of the flue by means of a manual optical pyrometer. However, this process is tedious, empirical and not very accurate. Above all, it provides no indication of the state of progress of coking, nor of the vertical temperature distribution in the cells. Some more up-to-date methods have been developed to improve temperature monitoring in coke ovens, but these do not seem to have yielded any permanent results. Marienau Experimental Station made a contribution in this field with the development, some twenty years ago, of a method of measuring the temperature in the charge by inserting thermocouples vertically through the oven mouth. This method is highly effective, but does not lend itself to monitoring current production.

We undertook to improve monitoring of the ovens' temperature regime by drawing up new representative temperature criteria, and by developing fixed measuring sensors and providing continuous recordings centralised in the control room.

We propose to describe in this paper three new methods which we have developed to determine:

a) the temperature level of the battery
b) horizontal and vertical temperature regulation
c) the degree of progress of coking.

II. MONITORING THE TEMPERATURE LEVEL OF THE BATTERY

II.1. Principle of the method

The principle of the method consists in measuring the temperature of the refractory bricks by means of a platinum/platinum-10% rhodium alloy thermocouple inserted through a hole drilled in the top of the ovens.

The major difficulty lies in drilling the holes in the dividing walls between two adjacent flues. For adequate sensitivity, the holes must be drilled to a depth of between 1.50 and 2.50 m, depending on the coking plant, so that the thermocouple's hot junction is situated at the top of the hairpin bend. To this end, we designed and constructed a centring device enabling the holes to be drilled with a high degree of accuracy (Figure 1).

At the present time, we can say that we have got this method of drilling down to a fine art.

We have drilled approximately ten holes to a depth of 1.60 m (as shown in Figure 2) without any major difficulty. It should also be mentioned that, by complying with our instructions, the operators of a modern coking plant were able to drill four holes measuring 2.5 m in a battery of tall ovens without encountering any difficulties.

II.2. Results obtained

The results obtained confirm the validity of this method of measurement. The temperatures measured in the crown of the ovens react well to inversion, to the charging cycle and to changes in tempo.

We attempted to establish some correspondence over a period of time between:

- the temperature measured by thermocouple
- the temperatures of flues situated at points along the thermocouple, measured by the traditional method of the optical pyrometer at the bottom of the flue
- the mean temperature of all the battery walls.

These different temperatures were found to parallel each other, as shown in Figure 3.

In conclusion, the thermocouples located in the refractory brick give a clear indication of the battery's temperature, and possess the advantage of providing a continuous measurement and registering trends more clearly. However, as it does not seem reasonable to install more than one or two thermocouples per wall, all the flues of the wall must be monitored in order to obtain some idea of horizontal temperature regulation.

III. MONITORING OF HORIZONTAL AND VERTICAL TEMPERATURE REGULATION

III.1. Principle of the method

The principle of the method is based on monitoring the side of the cake during discharging of the ovens, by means of an infrared pyrometer installed in a fixed position on the metal frame of the coke guide. It is possible to position two or three sensors at different levels in order to determine the vertical temperature distribution.

In order to test the effectiveness of this method, we carried out several series of tests at various French coking plant, and examined in detail the accuracy and reliability of the infrared sensors.

The actual development of the infrared pyrometer did not pose any problems. The system is easily adaptable to all the types of coke guide available in currently operating coking plant. Minor precautions must be taken to protect the pyrometer against dust and heat. The sensor is linked to a conventional potentiometric recorder installed in a sheltered position in the cab of the coke guide. The graphs obtained are serrated in appearance because of the fissured texture of the coke.

III.2. Results obtained

a) Monitoring of horizontal temperature regulation

We have, to date, compiled several hundred temperature charts using an infrared pyrometer, and the results are on the whole encouraging. These charts enabled us to draw the following three deductions:

- the infrared pyrometer positioned on the coke guide enables irregularities in carbonisation of the coke to be effectively pinpointed, and provides readings comparable to those of the

optical pyrometer.
- the readings obtained with the infrared pyrometer are much more sensitive than those of the optical pyrometer.
- the temperature graphs obtained in any one oven are reproducible.

Direct measurement of the coke's surface and the ability to position the pyrometer relatively far from the base plate seemed in the end to produce distinctly more representative results than measurement at the bottom of the flue, which is greatly affected by the design of the burner and the regulation of combustion.

Figure 4 shows an example of the parallelism between the readings taken by infrared sensor and traditional readings taken at the bottom of the flue. In this example, the pyrometer was positioned 2.2 m from the base plate.

Figure 5 illustrates the correlation between the coke's surface temperature measured during discharging of the ovens and the temperature measured at the bottom of the flue using an optical pyrometer. This correlation was established at a modern coking plant fired by rich gas, and is relatively sound. It, therefore, seems that measuring the temperature of the coke's surface during discharging may enable the temperature level of the walls to be assessed.

b) <u>Monitoring of vertical temperature regulation</u>

We completed the above measurements, carrying out simultaneous monitoring at two different levels.

Figure 6 shows some examples of recordings relating to a modern compound battery fired with lean gas and rich gas respectively. The pyrometers were positioned 2.2 m and 5.6 m from the base of the ovens.

The readings obtained give rise to some interesting observations. It seems that, with this type of oven, the two methods of firing result in different temperature distributions in the coke. When lean gas is used, the measurements taken at two levels in the cake produce very close results, with a slight displacement on the ram side. When rich gas is used, the curves relating to the two levels are, in contrast to the results obtained with lean gas, sharply staggered, revealing major over-carbonisation in the lower part of the cake, where the temperature recorded is largely in excess of 1000°C.

Parallel to this, we carried out temperature readings at the core of the coke, using thermocouples inserted into the charge through the oven mouth. This enabled us to prove that, in this type of battery, the vertical temperature graphs were highly satisfactory with regard to lean gas firing, but that they registered a delay in carbonisation at the top of the charge when rich gas was used.

Thus, the method of monitoring the surface of the coke during its discharge, by the use of infrared pyrometers, also enables the quality of vertical temperature regulation to be determined.

c) Application to monitoring of carbonisation of the coke

Figure 7 shows a graph of the rise in temperature which occurs during coking, one curve relating to the vicinity of the oven wall, and the other to the geometric centre of the charge. These two curves, which differ considerably during the first 17 hours, converge towards the end of coking and thereafter remain parallel. Thus, measurement of the coke's surface temperature by means of an infrared pyrometer can give an indication of the temperature at the centre of the charge, provided that a certain stage of carbonisation has been passed - in the above example, a surface temperature of 1000°C.

Figure 8 represents the correlation between the temperature measured at the surface of the coke and the readings carried out at the same level in the core of the coke, using thermocouples inserted through the oven mouths, at a well-regulated modern battery.

This correlation, which is good overall, tends to improve as the temperature increases, i.e. when the coke is over-carbonised. The results appear slightly more scattered at lower temperatures.

The readings obtained by infrared pyrometer can thus provide information regarding the degree of carbonisation of the coke, above a certain threshold which must be determined separately in each case.

IV. MONITORING THE DEGREE OF PROGRESS OF COKING

IV.1. Principle of the method

We have just seen how readings obtained by infrared pyrometer can, with certain precautions, provide information regarding the degree of carbonisation of the coke. Unfortunately, this information is not available until after discharging of the ovens, which is too late. What is needed, ideally, is a method of monitoring the progress of carbonisation from charging of the ovens until they are discharged, or at least during the last three or four hours of coking. Direct measurement of the temperature in the charge by means of thermocouples inserted through the oven mouths is an effective but cumbersome method, and does not lend itself to continuous monitoring. For this reason, we concerned ourselves with measuring the characteristics, in terms of temperature and composition, of the distillation gases.

The composition of the gases is determined by chromatography, using local samples taken from the base of the rising column. The temperature is monitored continuously, either at the base of the rising column or at its head, using a chromel/alumel thermocouple.

In our opinion, objective assessment of the degree of carbonisation of the coke resides in direct measurement of the coke's temperature in the cell, which should be used to standardize the frame of reference provided by the composition or temperature of the gases.

IV.2. Results obtained

a) based on the composition of the gases

Figure 9 shows the results of the gas analysis using a rising column at the end of coking. The curves relate to firing of a modern compound battery using lean and rich gas respectively.

A significant variation may be observed in the composition of the volatile matter remaining at the end of coking, and in particular a reduction in the hydrocarbon (methane, ethylene, benzene) content, the methane content being higher than that of the other constituents which are admittedly present in lesser quantities. Furthermore, these results permitted us to distinguish between the two methods of firing, greater quantities of methane, ethylene and benzene being released at a mean temperature and in an identical charge when rich gas is used. Inadequate carbonisation in the upper part of the cake when rich gas is used to fire this type of oven is responsible for these higher methane, ethylene and benzene contents.

Figure 10 consists of curves showing the development in composition of the gases in the rising column, not this time in relation to the mean temperature at the core of the coke, but rather in relation to the temperature in the top of the charge. When lean gas is used for firing, the temperature in the top of the charge is of the same order as the mean temperature. Contrastingly, when rich gas is used, the temperature in the top of the charge is much lower than the mean temperature. This being the case, the points corresponding to the two methods of firing are remarkably well-aligned. This shows that the composition of the gases in the rising column is directly related to the temperature of the less-carbonised zones. It, therefore, seems possible to draw up an end-of-coking criterion, based on the limit values of the hydrocarbon content. The acquisition of these limit values means that all parts of the cake are heated to a specific temperature, whatever the quality of vertical temperature regulation.

b) <u>based on the gases' temperature</u>

Figure 11 shows a graph of the temperature variation in the rising column (approx. 50 cm from the base of the column) and at the head of the column. The readings taken at the head of the column show a greater temperature reduction at the end of coking than observed actually in the rising column.

The readings taken at the head of the column may thus be preferred because of their greater sensitivity. Figure 12 illustrates, by way of example, the principle of a graphic method of determining an end-of-coking criterion, based on measuring the distillation gases' temperature and applicable to all batteries provided prior standardisation of results is carried out.

The time scale was extended in comparison to Figure 11, to give a better picture of the descending section of the curves relating to the distillation gases' temperature at the end of coking. On this, we have superimposed the curve showing the mean temperature at the core of the coke, recorded through the oven mouths. If we stipulate, for example, that the coke should be discharged from the ovens when the mean temperature reaches 950°C, we can then determine the corresponding temperatures in the rising column (point C) and at the head of the column (point T) by means of a simple graphic construction. The temperatures obtained are 785°C and 455°C respectively.

The same graphic construction may be repeated, paying particular attention to the retarded carbonisation zones rather than the mean temperature of the charge. But, judging by the results obtained to date, it seems that the temperature of the distillation gases is more closely related to the mean temperature of the coke than to the lesser-carbonised zones.

V. CONCLUSIONS

Without any major technical difficulties, we have succeeded in developing three new methods of measurement, suitable for continuous monitoring and whose recordings can be centralised in the control room.

Measurement of the temperature of the refractory brick by means of thermocouples inserted into the crown of the oven is a good method of recording the temperature level of the battery, and can replace to good advantage the conventional readings obtained by means of an optical pyrometer.

Measurement of the coke's temperature during discharging of the ovens provides, on one hand, an accurate picture of horizontal temperature regulation and, on the other hand, enables the quality of vertical temperature regulation to be determined. Additionally, the results obtained by this method appear more representative and more sensitive than those obtained using a conventional optical pyrometer. Furthermore, this method of measurement can provide information regarding the degree of carbonisation of the coke, provided that the results have previously been standardised.

Measurement of the distillation gases' characteristics, temperature and composition, using a rising column enables us to determine the degree of carbonisation of the coke.

What remains to be done is to examine the problems posed by the behaviour and reliability of the sensors in the long term, and to solve certain problems in transmitting the data to a central computer which will process the information in real time.

Figure 1 : Centring device for drilling a hole in the top of the oven (Marienau coking plant)

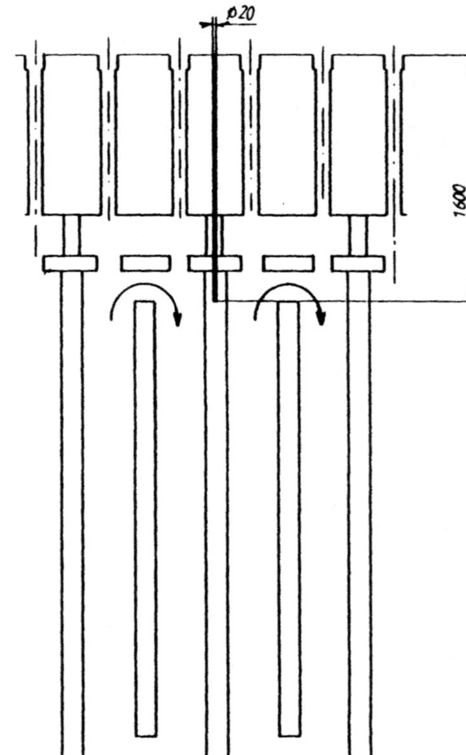

Figure 2 : Positioning of the thermocouple in the dividing wall between two flues (Marienau coking plant)

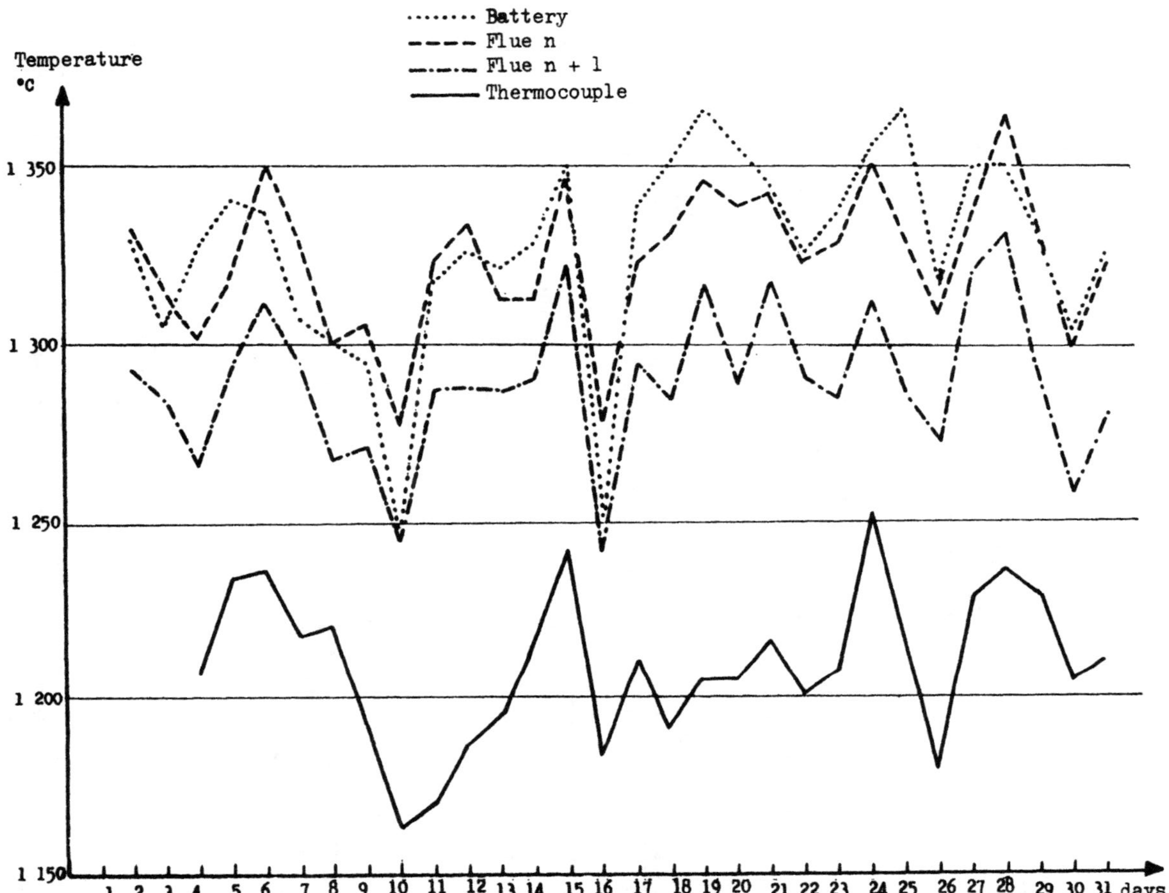

Figure 3 : Correspondence between the temperatures measured by thermocouple, the temperatures of adjacent flues and the mean temperature of all the walls (Dunkirk coking plant)

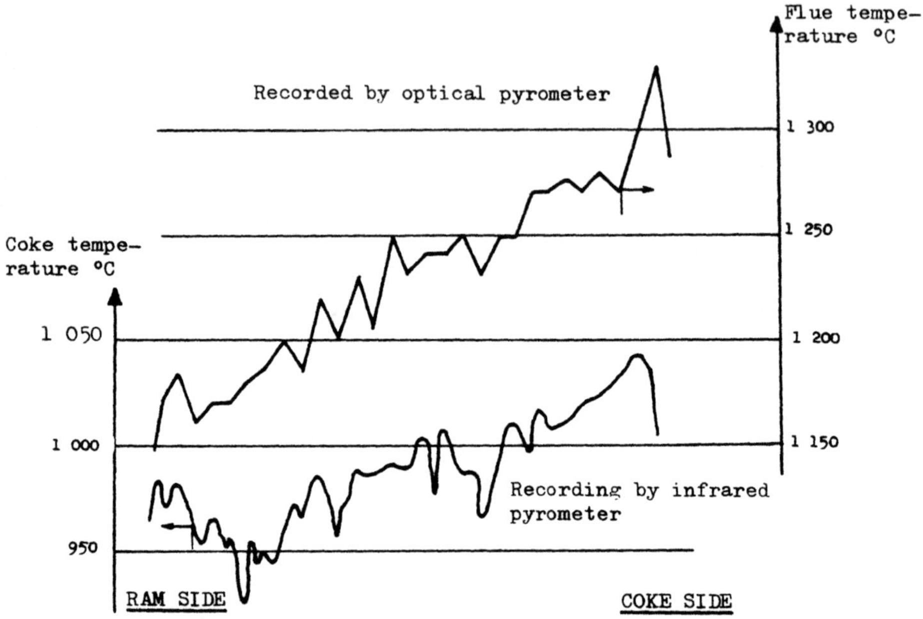

Figure 4 : Temperature charts of the flues and the side surface of the cake recorded at a modern battery (Fos-sur-Mer coking plant)

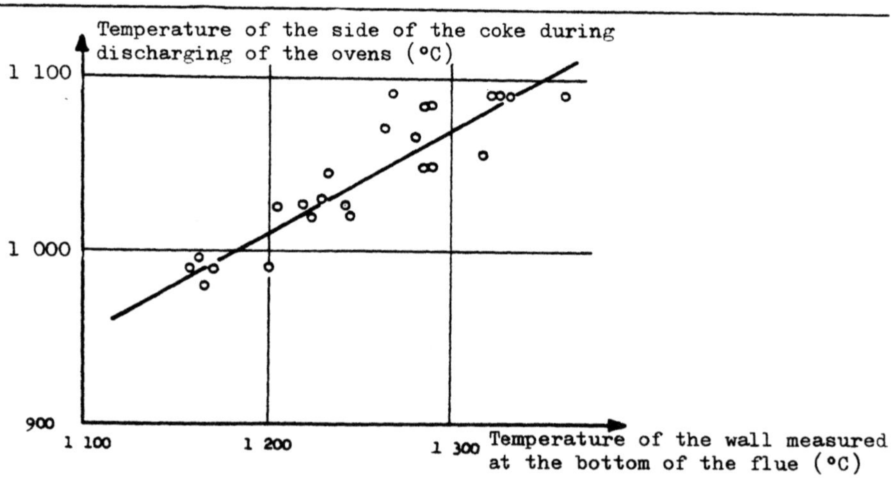

Figure 5 : Correlation between the temperature of the side of the coke during discharging of the ovens and the temperature of the wall measured at the bottom of the flue

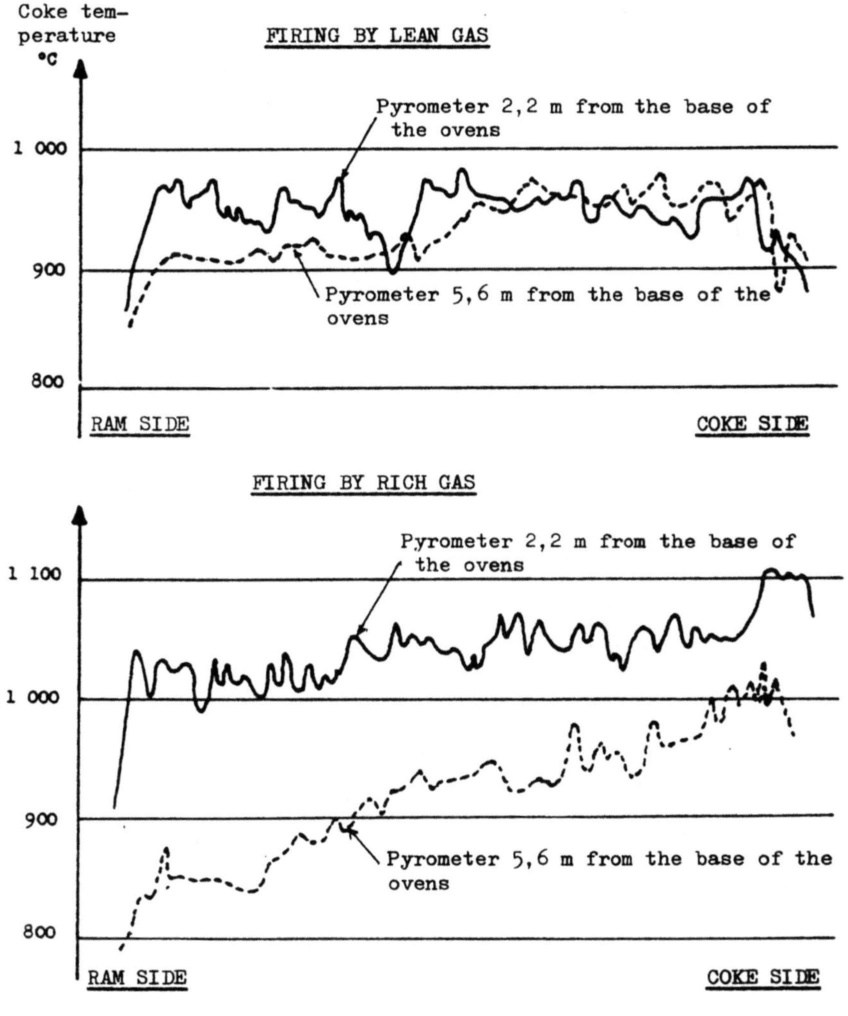

Figure 6 : Chart of the coke's surface temperature recorded by infrared pyrometers positioned at two levels (Fos-sur-Mer coking plant)

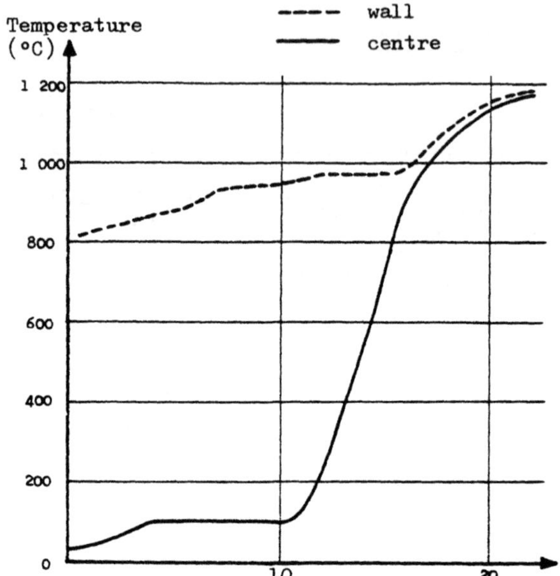

Figure 7 : Curves showing the rise in temperature in the vicinity of the wall and at the eometric centre of the charge (Marienau coking plant)

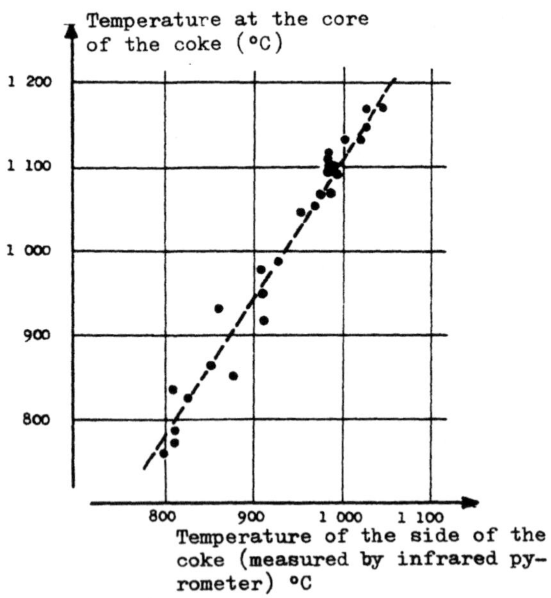

Figure 8 : Correlation between the temperature at the core of the coke and the coke's surface temperature, measured during discharging of the ovens (Fos-sur-Mer coking plant)

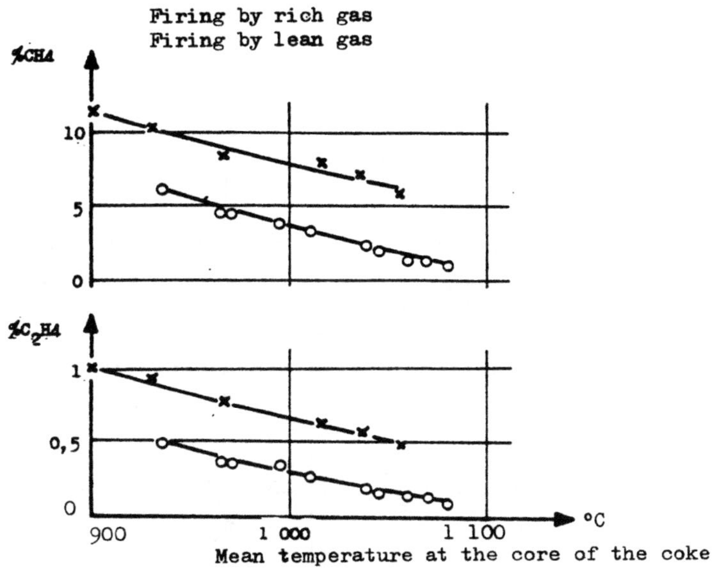

Figure 9 : Trend of gases' composition in the rising column in relation to the mean temperature at the core of the coke (Fos-sur-Mer coking plant)

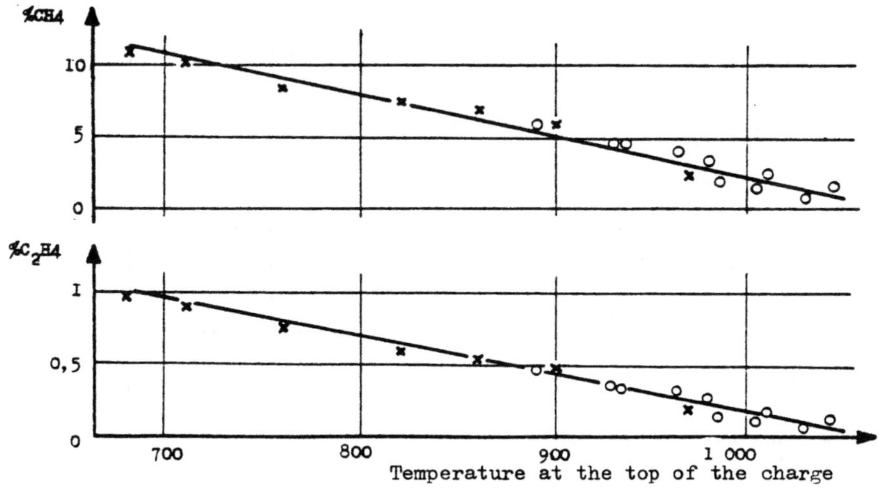

Figure 10 : Trend of gases' composition in the rising column in relation to the temperature in the core at the top of the charge (Fos-sur-Mer coking plant)

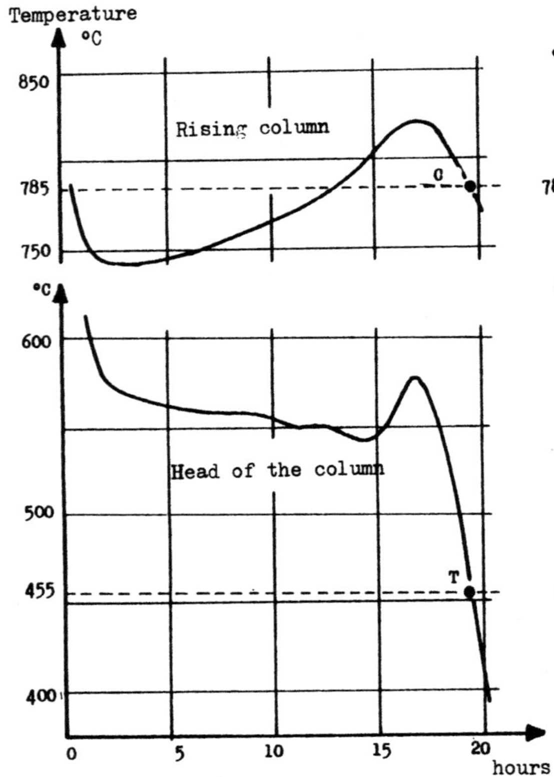

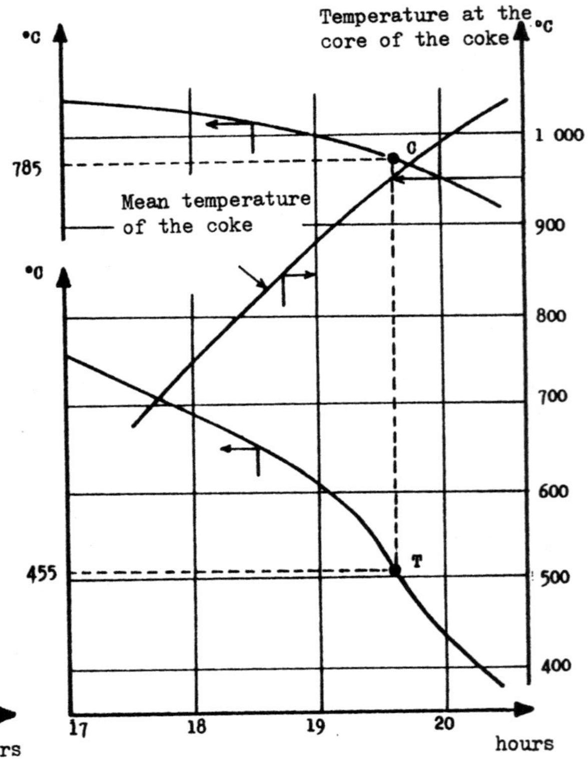

<u>Figure 11</u> : Curves showing development of the distillation gases' temperature (Marienau coking plant)

<u>Figure 12</u> : Attempt to establish an end-of-coking criterion on the basis of the distillation gases' temperature

TESTING OF THE CODECO SYSTEM FOR THE PROGRAMMED HEATING OF COKE OVENS

D. Habermehl

Bergbau-Forschung GmbH

Summary

The first full scale technical application of research into the CODECO programmed heating system as carried out at the now-defunct Emil Experimental Coking plant was in the construction of the new Prosper Experimental Coking plant at Bergbau-Forschung GmbH.

The programmed heating system was successfully tested over operating period of 18, 16 and 14 h which demonstrated that typical coke oven temperature levels could be considerably reduced as compared with conventional oven designs and heating arrangements. Heating by programme also led to savings of as much as 6.4% firing related energy. The CODECO system had no effect on coke products in terms of volume, type or composition.

INTRODUCTION

A report to the symposium on Coking Technology held here in Luxembourg in May 1975 outlined current work on the development of the CODECO system (1). The work had been brought to a conclusion by the time the Emil plant closed down at the end of 1975 (2) and its immediate technical outcome took the form of three full scale test ovens at the new Prosper plant (3). There follows a description of the construction and testing of this heating system.

1. THE CODECO SYSTEM

Investigations at Emil (4-6) had revealed the following:

- heat supply and demand are best balanced by controlling the supply rate;
- it is pointless trying to control heat supply by temperature feedback because of the high response inertia of the coke oven system as such;
- admittedly the unwanted increase in end-product temperature which occurs towards the end of the carbonization period can be overcome;
- but at the same time consideration must be given to the storage capacity of the firebricking in order to have sufficient heat available for starting up the next charge;
- the sluggish response of the heat transfer system must be taken into account when setting up the heating programme, and
- a relatively coarse system of block phasing is every bit as effective as meticulous adjustment of heat supply and demand.

In the course of the work a definite method evolved for lining up the heat phase gradations so that a heating programme is made up as follows:

- two to three stages with fairly long interrupt periods, each stage being unalterable as such;
- a gas-off mode in which the entire heating system is for all practical purposes cut off and no gas is fired.

Using an optimized programme of this kind it becomes possible to push completed burns at coke end temperatures below 1000°C after an 18 hour coking period and in this way up to 10% of firing costs.

The technical details of this programmed heating system were worked out by Bergbau-Forschung GmbH in conjunction with Dr. C. Otto GmbH and Carl Still GmbH & Co. KG and subsequently given the name of CODECO. In this system, a battery is normally broken down into groups of five to eight ovens roughly in the same stage of carbonisation and which are pushed and filled in close sequence. Gas and air volumes are varied as within any one group (Figure 1). This can be achieved by altering the gas and air pressure, typically compressing the air in a blower and forcing it in through the pipework system as depicted in Figures 2 and 3. The air switch cocks will be seen in Figure 2, whilst Figure 3 shows the sealed air inputs leading into the regenerator. Stack draught in the waste heat valve can be controlled, individual regulators being fitted to each waste heat well for this purpose. The pause following on a heating reversal can be varied by programme as and when required.

The requisite heat is supplied in different stages and thus better adapted to the charge with progressive carbonising. Up to four heating stages (including the gas-off mode) are undertaken by a freely programmed control system. Apart from the actual controller, this system includes a programmer for the input, modification and testing of programmes. The controller actuates the reference input transducers for all regulators and controls (air, gas and waste heat). The transducers can be set per heating stage right over the control range.

In contrast to the normal fixed wiring hook-ups, the entire control function here is computerised. Manual overrides are available to cover electronic malfunctions.

The heating equipment itself can be adapted for forced-air routines. The non-regulated initial pressure can be raised by means of the blower and the regulated pressure, and thus the volume supply, is held constant by a pneumatic regulator as appropriate to the particular heating stage. Fixed-point piping feeds in the combustion air to the individual regenerators, and ball shutoff cocks are used to control this during heat switching. Safety equipment includes an air flap valve which opens automatically in response to fan failure and so ensures that the heating paths are adequately purged.

An additional cutoff cock is built in between the gas switch and hand shutoff cocks in each injection line, and is actuated via the controller, but closed during the gas-off mode. Air supply can be choked down to minimum flush by means of the regulator. The chokes below the off-heat valves are closed to the amount required without in any way affecting normal reversal action.

2. TEST EQUIPMENT AND OPERATIONS

At the Prosper experimental plant (3), the CODECO heating system has been installed in three high-performance silica lined test ovens, commissioned in 1976 along with the 36 ovens making up battery VII. All ovens were furnished with 80 mm thin style stretchers right the way up, whilst the test ovens for their part were supplemented by regenerator checkers providing an extensive heat exchange surface.

Coking trials with these ovens (7) have indicated that their operating periods compare with those of conventional ovens, for an appreciably lower temperature level. The energy savings, or alternatively increased output, made possible by this will at once be obvious.

Coking tests to check out the programmed heating system were divided first of all into three series of experiments to be run off at times of 18, 16 and 14 h respectively. These were followed by additional trials at the 18 h level using a modified heating programme aimed at achieving further energy economies by reducing the coke end temperatures as far as possible.

During the coking tests, the oven group was operated on a block principle as distinct from the conventional pushing cycle so that all three charges were at roughly the same stage of carbonisation. This is the only way in which energy supply could be programmed to suit the requirements of all three charges. All measurements conducive to volume and energy balancing were carried out as a matter of course during these tests.

3. FIXING THE HEATING PROGRAMMES

Investigations commenced with a 14 h operating period block type programme based on results of earlier programme trials (6). This programme was subsequently applied to the 16 and 18 h periods.

As will be seen from Figure 4, the heating programme is devised as follows: it starts out from a heating stage to cover the high initial heat demand, corresponding to about 170% of the normal gas flow on a steady firing schedule: this is sustained over about one third of the coking period. To prevent overheating the firebrick, especially in short runs, this phase is followed by a one-hour heating pause. A second heating stage ensues, supplying something like 110% of the reference gas volume and once again for a further third of the operating period. This II stage gives way to a break of 3-4 h to achieve the intended low end temperature of the coke product. Heating stage III which intervenes one hour before pushing is similarly rated for 110% gas volume and is responsible for raising the wall temperature so as to provide a heat store for the next charge of cold wet coal, once again associated with the high 170% gas initial heating stage I.

4. TEST RESULTS

4.1. Oven preparation

The average tuyere temperature cannot be used as a characteristic boundary condition to describe operating conditions in programmed heating as it is governed by the programme variables and has, therefore, to be determined separately for each of the three stages. The operating (coking) time was, therefore, used in its place.

Another difference in the measured tuyere temperatures compared with other types of oven results from the 600 mm burner tubes fitted to the headwall tuyeres so as to improve temperature distribution over the height of the heating flue wall. Values as measured are about 50 K lower than with comparable type of oven devoid of the burner fitment (7).

As far as the different temperature indices of oven operation are concerned, the same basic tendencies were noted with the changed oven schedule as had been found when operating conventional ovens on a steady-firing principle (7). One essential difference which did arise, however, was the appreciably lower general level of all the temperature indices achieved when the programme-responsive tuyere temperatures in the individual stages are disregarded.

The reduction in the characteristic temperatures becomes more marked with any shortening of the operating period. Thus for wall and oven centre temperatures (Figure 5) taken by profile measurements within the charge, there is a clear reduction of 70 or 30 to 70 K directly attributable to programmed heating.

When comparing these results with those from conventional ovens, the effect of the relatively thinner stretcher bricks is also evident in that the two temperature values for a comparable operating period of 18 h are as much as 100 K lower than in normal types of oven. The optically measured wall temperatures at the base and crest of the charge are similarly affected, as might be expected.

The average gas space temperature can be slightly lowered around 20 K using programmed heating (Figure 5). In comparison with conventional ovens there is, however, a reduction of 60 K for an 18 h operating period chiefly due to the lower heating flue temperature brought about by the thinner stretcher design.

The reduction, in places appreciable, of the characteristic oven temperature for a given coking period caused by the special design of the prototype high-performance silica ovens necessarily affects the heat balance during carbonisation. Comparative investigations using this new type of oven have shown that the programmed heating alone saves 6% of the heat consumption. A comparison with earlier heat balances using conventional coke ovens does, however, pose problems, not only from the altered raw materials basis.

Nevertheless an additional saving of between 7 and 7.5% energy can be estimated for a given coking time of 18 h as a result of reduced flue temperature by the use of the thinner stretcher bricks and reduction in off-gas temperature, as also improved regenerator checkerwork.

The heat balance of the centre oven in programmed heating is shown as a heat flow diagram in Figure 6 for the sake of clarity.

The familiar increase in heat consumption with rising flue temperature is naturally also present with the programmed oven. It stems from an increase in sensible heat in the coking products and the off-gass temperature, combined with a possible kinetically motivated fall-off in reaction heat.

Figure 7 shows the effect of the different heating systems on surface and off-gas losses. It demonstrates that where heating is programmed the off-gas losses are slightly higher as a result of firing pauses than in the case of constantly maintained heating. In comparison with conventional ovens, on the other hand, the off-gas losses are much lower which is due in large part to the superior heat exchange facilities offered by the new design of checkers.

The surface losses drop back slightly with any reduction in coking time despite increased radiation losses as both the amount of coal put through per unit time and the imported firing heat rise more powerfully than the irradiated heat volume. They are clearly less affected by aspects of heating technology such as programme control.

The investigations carried out subsequent to the main test, using a modified heating programme, have shown that with the 18 h time, the coke end temperatures can be held to 1000°C throughout, and that carbonisation is complete under these conditions.

The forced energy input imposed by programme heating at the start of the coking period provides a sharp lift in charge temperature in the first two thirds of the coking period. In contrast to constant heating, therefore, the moisture evaporation phase is shortened to the extent of about 8%. Coking progress, defined as the temperature rise in the charge between 350 and 550°C on a time basis, is not significantly affected.

The relatively long off-heat phase in the final third of the coking period reduces any further increase in the charge temperature to an extent which gives rise to the reduced coke end temperature detailed above. A study of shrinkage suggests satisfactory levels even in the case of shorter runs carried out under high-performance conditions. Programmed heating would not appear to have any effect on shrinkage.

4.2. Coking products

It is not possible to compare the coke indices with those achieved in conventional ovens as the charge consistency has changed from those used in earlier tests, and also because the investigations to-day are carried out with stabilised cokes in the interests of comparability with plant operating indices. The size consist tends to shift to the smaller coke sizes with the shorter operating period, and this together with the reduction in coke strength is in principle also found with high-performance coking (8). No significant relationship was found between programmed heating and coke qualityß There was a secondary improvement in size uniformity as a result of a diminished proportion of the plus 80 mm coarse fraction.

As far as by-products are concerned, the effect of shortened operating as already evidenced in horizontal ovens, was confirmed with the further reduction in operating time (8). Volume, type and composition of the by-products are not noticeably affected by heat programming in the high-performance oven. This is to be expected in view of the small degree of influence exerted on the coking operation by temperature changes in the range of temperatures which are of pyrolytic relevance. There is a slight alteration in the release timing of the gasification products as a result of programme control: thermal cracking conditions differ as between the stretcher wall and coke charge in the space traversed by the gas. It is the hydrogen development which is most powerfully affected in this context.

5. OUTLOOK

In conclusion it must be emphasised that ovens heated by the CODECO system are fired exclusively by rich gas. Suitability for dilute gas firing is to be tested shortly on ten ovens in a commercial battery furnished with a CODECO heating system.

Tests are in train at the experimental plant at the time of writing aimed at discovering to what extent the combustion air blower is essential, in other words whether combustion can be supported by natural draught. The suggestions are that this may well be possible under certain circumstances.

We would like to express gratitude to the Energy Directorate of the CEC for financing our project development.

BIBLIOGRAPHY

(1) ROHDE, W., STALHERM, D. and BECK K-G:
Development of a high-performance silica lined coke oven; paper at Information Symposium on Coking Techniques of CEC, Luxembourg 6/7.5.75. pp. 75-94.

(2) BECK, K-G:
The Emil Experimental Coking Plant. Glückauf 111 (1975) pp. 1125-1127.

(3) HABERMEHL, D. et al:
The new Prosper experimental coking plant. Glückauf 114 (1978) pp. 643-647.

(4) ROHDE, W., HABERMEHL, D. and MEYER, C:
CODECO - a new technique for controlled coke oven underfiring. Paper at 106th AIME annual meeting, Atlanta 1977.

(5) ROHDE, W., SIMON, G. and BECK, K-G:
Energy demand and supply of blast furnace coke production. Paper at Proc. Ironmaking Conf. 38 (1979) pp. 46-51.

(6) HABERMEHL, D., ROHDE, W. and BECK, K-G:
The development of the CODECO programmed heating system. Glückauf 117 (1981) pp. 17-23.

(7) HABERMEHL, D., ROHDE, W. and BECK, K-G:
The effect of coking times on the heat balance in a high performance oven charged with wet coal. Glückauf 116 (1980) pp. 168-74, 186.

(8) HABERMEHL, D., KUCHARZYK, W. and ROHDE, W:
The effect of short coking times on coke quality and quantity with wet charge coal. Glückauf-Forsch.Fefte 41 (1980) pp. 121-125.

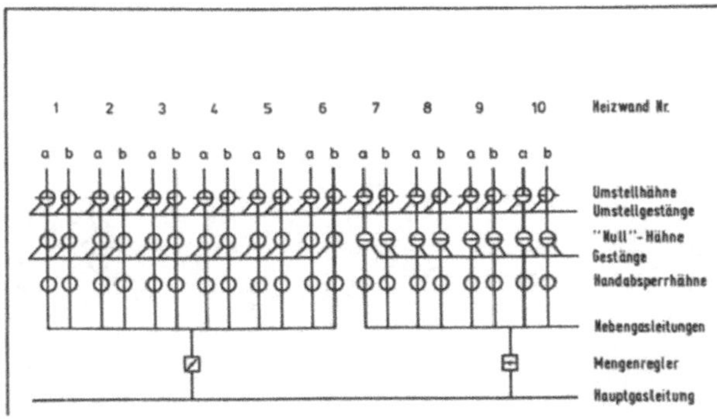

Heizwand	**heating wall**
Umstellhähne	**switch cocks**
Umstellgestänge	**switch linkage**
Null-Hähne	gas-off cocks
Handabsperrhähne	manual shutoff cocks
Nebengasleitungen	ancillary gas lines
Mengenregler	volume regulator
Hauptgasleitung	gas main

Figure 1. Group gas supplies.

Figure 2. Switchcocks for the CODECO programmed heating system.

Figure 3. Sealed air inputs for the CODECO system.

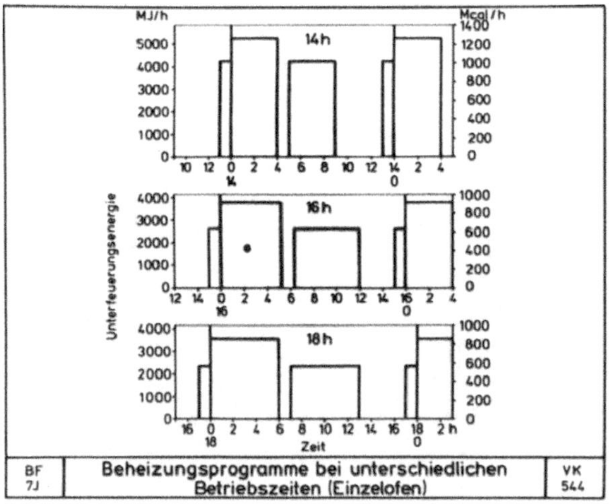

Unterfeuerungsenergie — firing energy
Zeit — time

Figure 4. Heating programmes for different operating times (single ovens).

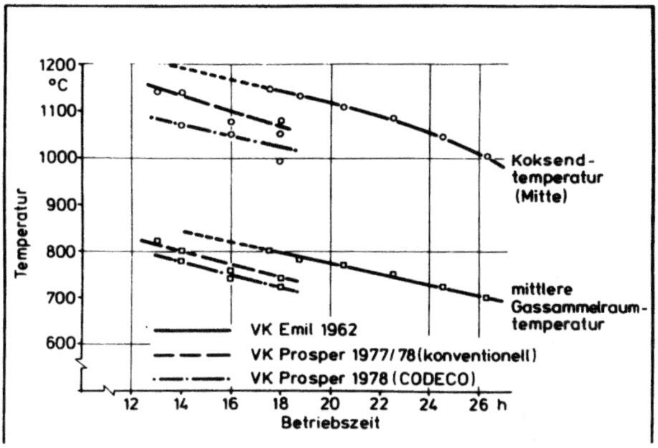

Temperatur — temperature
Koksendtemperatur (Mitte) — coke end temperature (average)
mittlere Gassamemelraumtemp — average gas space temperature

Figure 5. Coke end temperatures and gas space temperatures for different heating systems.

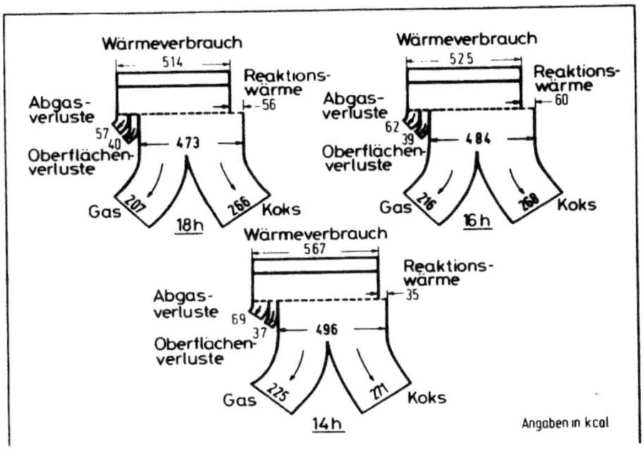

Wärmeverbrauch — heat consumption
Abgasverluste — off gas losses
Reaktionswärme — reaction heat
Oberflächenverluste — surface losses
Koks — coke

Figure 6. Heat flow diagram for coking coal with 10% V.M. over various coking periods.

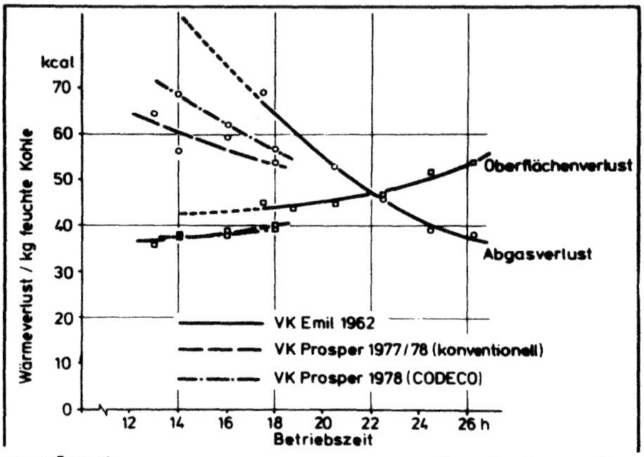

Wärmeverlust	heat loss/kg moist coal
Oberflächenverlust	surface losses
Abgasverlust	off-gas losses
Betriebszeit	operating time

Figure 7. Surface and off-gas losses for different heating systems.

THIRD TECHNICAL SESSION

Chairman: J. VAN LANGEN, Head of Process Technology Department, Estel Hoogovens BV

- Rapid devolatilization in the coking process
  L. PALUMBO and P. ZANCHI, Centro Sperimentale Metallurgico

- Partial repair of damaged coke chamber walls
  N.J.W. THIJSSEN and P.G.J. BROERSEN, Research and Development Laboratory Estel Hoogovens BV

- Summary of round table "A" Coke and the blast furnace
  G. SIRONI, Centro Sperimentale Metallurgico

- Summary of round table "B" Pollution problems
  J.P. GRAHAM, British Carbonization Research Association

- Closing address
  A. DE GREEF, Head of Division "Technologies, Synthetic Fuels, Research and Development", Coal Directorate, Directorate-General for Energy, Commission of the European Communities.

THIRD TECHNICAL SESSION

Chairman: J. VAN LANGEN

SUMMARY OF THE DISCUSSION

Papers:

4) Rapid devolatisation in the coking process, by L. Palumbo and P. Zanchi.
5) Partial repair of damaged coke chamber walls, by P. Broersen and N. Thijssen.

The cohesion of the subjects with difficulty can be found in the fact, that both consider the effect of temperature fields, one on the coal mixture to be coked, the other on the behaviour of the refractory construction of the coke oven.
A high temperature gradient increases the coking reaction, but has also a marked effect on the quality of coke and gas produced.
The necessity of studying the repair methods of the refractory construction of the oven is caused by the nowadays often irregular operation of the batteries, as discipline has been replaced by democracy in plant operation.

Discussion.

Mr. F. Orywal of Bergbau Forschung has three questions for Mr. Palumbo, who subsequently provided the answers.

1) What type of refractory has been used for the experimental coke oven.
   As the temperature field to be examined reaches till 1600°C alumina has been chosen for the construction of the oven.

2) For the hot coke quality test Mr. Orywal believes that CSM started to use a refractory clad drum, which now seems to have been replaced by a steel drum.
   However, CSM has never used a refractory drum. The steel drum consists of centrifugally cast steel with a high chrome-nickel content.

3) According to Bergbau Forschung's experience the composition of the gas produced in an experimental furnace is not comparable to that obtained in a coke battery.
   What are the experiences at CSM?

Mr. Palumbo confirms that the gases expelled from an experimental oven do not have the same time for cracking as in a production oven. He believes however, that the tendances indicated by the experiment will qualitatively hold true also in praxis.

Mr. Simpson of BSC Port Talbot asks Mr. Thijssen about the praxis to charge coal behind a stopper wall of an oven in partial repair.
Does this mean that the stopper wall must be gastight or is it necessary to wait with the repair work until the gas production has died down? After the repair how is the stopper wall removed?
Mr. Thijssen answers that charging of coal behind the stopper wall has been the praxis already before the study of the repair method began. No ill effects have ever been found, so there was no reason to change on this point.
After repair some bricks at the bottom of the stopper wall are taken out, so that the stopper wall gets loose. It is then pushed out with the coke.

RAPID DEVOLATILIZATION IN THE COKING PROCESS

L. Palumbo and P. Zanchi

Centro Sperimentale Metallurgico

1. INTRODUCTION

It is well known that the characteristics of the input blend and the operating conditions of a coking plant have a considerable influence on the quality of the metallurgical coke produced.

It is also known that the possible variations in the operating conditions of a coking plant can affect the temperature of the walls and the stabilization time.

It is, therefore, advisable to study the influence of these parameters on the quality of the coke with the help of experimental ovens, which, compared with an industrial coking plant, allow such operational parameters to be varied more easily and more widely.

Research has accordingly been carried out at the CSM (Centro Sperimentale Metallurgico = Metallurgical Test Centre), with support from Italsider, with the objective of studying the influence of various devolatilization rates and of two thermal stabilization levels on the properties of the products of coking, using input blends with volatile matter contents similar to those found in industrial situations.

2. RESEARCH EQUIPMENT

2.1. Experimental coke oven

The research was carried out using the CSM's experimental coke oven.

This oven (Figure 1) is electrically heated and is fitted with an adjustable wall.

The main specifications of the oven are as follows:

- width of chamber 0.45 m
- useful height of chamber 0.91 m
- useful length of chamber 0.91 m
- weight of wet charge of coal 300 - 350 kg.

The oven can accommodate two different heating systems: the first consists of twelve silicon carbide resistors (six for each wall) which allow a maximum temperature of 1100°C to be reached on the internal wall of the chamber.

The second system consists of twelve suitably mounted and protected molybdenum resistors which allow a maximum wall temperature of 1350°C to be reached.

The coal is gravity charged from the top of oven, whilst it is discharged horizontally by means of a special machine which pushes the coke into a box where it is dry-cooled.

The gases released during devolatilization, once they have been cooled by washing with water, are passed through a special valve which allows the pressure within the oven to be regulated to approx. 20 to 30 Pa (2-3 mm W.G.) and are finally exited to the stack.

Thermocouples and special probes introduced into the charge record changes in temperature and in the pressure exerted on the coke cake.

2.2. Auxiliary equipment

The coke oven itself is supplemented by a whole range of auxiliary equipment for preparing the charge and analysing the coke produced.

This equipment is of a conventional nature, with the exception of the equipment for measuring the characteristics of the hot coke in a reactive atmosphere; this piece of equipment was designed and constructed at the CSM.

2.2.1. Equipment for tests on hot coke

The prototype equipment constructed at the CSM (Figure 2) allows the behaviour of the coke under dynamic and abrasive stress, in a gaseous atmosphere, to be studied up to a temperature of 1100°C; coke of typical size consists is used.

The test chamber is a stainless steel cylinder, with an internal diameter of approx. 280 mm and a useful length of 900 mm, into which the coke is placed.

The chamber is sealed with two suitable stoppers, in which a tubular opening is made so as to allow gases to be fed into and released from the chamber during tests.

The chamber is heated from the outside by the combustion gases of two burners, up to a test temperature of 1100°C. It can also be rotated (at 10 rpm) together with the combustion chamber.

For the purposes of this research, the hot coke tests were performed as follows: the coke sample, which weighed 7.5 kg and had a size consist in the 40 to 60 mm range, was placed in the test chamber and heated in a nitrogen atmosphere; when a temperature of 1100°C had been reached, the flow of nitrogen was stopped and carbon dioxide was introduced (at a rate of 50 Ndm^3/min) whilst rotating the chamber through 700 revolutions.

At the end of the test, the coke was transferred into a suitable container and dry-cooled.

The cooled coke was screened using 20 and 10 mm diameter round-hole sieves; in this way, the following indices were obtained:

CF_2 = Percentage over 20 mm after hot coke test

CF_1 = Percentage under 10 mm after hot coke test, including losses

$$CK = \frac{CF_2}{CF_1}$$

3. TEST METHOD

The research work was done using three blends with a volatile matter content of approx. 23, 25 and 27% respectively.

The 25% v.m. blend was taken direct from the feed belt at

an Italsider works.

The blends with other v.m. contents were obtained by adding appropriate quantities of low and high v.m. coals to the 25% v.m. blend, so as to adjust the percentage of these coals in the blend.

The main chemical properties, plastometric behaviour (measured using a Brabender plastograph) and size consist characteristics of the three blends concerned are shown in Table I.

Each blend was devolatilized at the following average chamber-wall temperatures: 1050°C, 1100°C and 1300°C.

In order to stabilize the time of the end of devolatilization, the following principle was adopted: by means of a suitable thermocouple, the time taken for the temperature in the middle of the charge to reach equilibrium with that of the walls (a temperature increase of less than 5°C/0.5 hrs) was measured, and the coke was pushed 0.5 hours and 1.5 hours after this equilibrium was reached.

The total number of charges pushed was, therefore, 48, corresponding to 3 blends, 4 wall temperature levels and 2 thermal stabilization levels (test carried out twice for each combination).

The coke produced, once dry-cooled, was screened and the fraction over 75 mm in diameter was crushed by a jaw crusher.

The micum and Irsid tests were carried out on representative samples of both the untreated coke and the crushed coke.

In addition, the crushed coke was also subjected to a hot-coke test in a reactive atmosphere.

For each charge, a representative sample of the coke oven liquor was also taken and analysed in order to determine what pollutants it contained.

4. SUMMARY OF MAIN TEST RESULTS

The main results of the research work are summarized in

Tables II and III and in Figures 3 to 9.

Table II lists the main characteristics of the devolatilization process for the four wall temperature levels tested.

From this Table, it will be seen what a wide range of test conditions the use of the CSM pilot plant made possible; by increasing the wall temperature from 1050°C to 1300°C, the average devolatilization time before reaching the thermal equilibrium defined above decreased from 16.11 hours to 12.84 hours, a 20.3% reduction.

Accordingly, the devolatilization rates and the specific yields per hour, defined by the following ratios:

$$\frac{\text{width of chamber}}{\text{devolatilization time}} \qquad \left[\frac{mm}{h}\right]$$

$$\frac{\text{kg of coal charge (dry)}}{\text{useful volume of chamber x devolatilization time}} \qquad \left[\frac{kg}{m^3h}\right]$$

rose respectively from 27.9 to 35.0 mm/h and from 46.5 to 58.3 kg/m^3h, an increase of 25.4%.

To obtain the discharging times, the stabilization times tested are added to the average devolatilization times before thermal equilibrium is reached. The average temperatures reached in the middle of the charge before discharging were also increased under the test conditions adopted, from 1016°C to 1187°C (+ 16.8%).

Corresponding to the wide range of test conditions, there were wide variations in the characteristics of both the coke and the coke oven liquor.

In order to identify the influence of each of the separate variables adopted in the tests, i.e.:

- wall temperature,
- stabilization temperature, and
- input blend

on the characteristics of the products of coking, the data obtained was interpreted by means of analysis of variance.

The results of the analysis of variance are shown in

Table III, in which the statistical significance is indicated by one, two or three asterisks, corresponding to a level of probability of 95, 99 and 99.9% respectively.

As will be seen from the Table, the influence of the stabilization time was only evident in the case of the CSM hot-coke test.

The stabilization brought about an improvement in the properties of the coke, but the resultant changes were only slight.

In practice, only the different blends had any influence on the properties of the coke produced, and even in this case, the variations, although statistically significant, were slight.

The blends with v.m. contents between 23 and 27% yielded a coke with inferior cohesion properties (lower I_{20} and greater M_{10} and I_{10}) when compared with coke produced from the 25% v.m. blend.

However, one independent variable influenced all the characteristics of the products - both coke and coke oven liquor - namely the oven-wall temperature.

We will, therefore, restrict our more detailed discussion to this variable, which influenced the products of coking under the test conditions adopted either by causing a significant increase in the devolatilization rate or by raising the final temperature of the coke.

In order to show more clearly the relative importance of the influence of this independent variable on each individual characteristic, we have prepared histograms for the various treatment media (see Figures 3 to 9).

In these histograms, the four levels of test temperature are plotted on the x-axis, whilst the averages for each temperature level for the various dependent variables are plotted on the y-axis; in each case, the overall average of each individual characteristic is taken to be 100.

Figure 3 shows the middle size ranges of the untreated and of the crushed coke; it will be seen that, as predicted, there is a sharp drop in the middle size range of the untreated coke, whilst the middle size range of the crushed coke is remarkably stable.

Thus it appears that subjecting the coke produced to mechanical stabilization by crushing it does not result in any deterioration in the coke's middle size range when the devolatilization temperature rises.

Figures 4 and 5 show the mechanical properties of the cold coke as measured by the micum and Irsid tests. As predicted, the M_{40} index drops sharply because of the increased fissuring of the coke caused by the rise in the devolatilization rate.

The increase in the devolatilization rate and in the final temperature also, however, caused a marked improvement in the cohesive properties of the coke, as demonstrated by the drop in the M_{10} and I_{10} values.

The I_{20} index also showed a statistically significant but slight improvement.

There were important variations in the results of the CSM hot-coke and reactive atmosphere tests; whilst there was only a slight variation in the CF_2 index (the fraction over 20 mm), there was a very marked change in the CF_1 index (the fraction over 10 mm) and in the CK index (the $\frac{CF_2}{CF_1}$ ratio).

The CSM hot-coke test also (Figure 6) showed up an improvement in the coke quality as the wall temperature rose from 1050°C to 1200°C (57.6% increase in the CK index), followed by a marked deterioration in the next temperature stage between 1200°C and 1300°C (13.9% drop in the CK index).

The cold-coke tests did not indicate a similar pattern in the coke characteristics, with the exception of a slight rise (+ 1.8%) in the I_{20} index as the wall temperature increased from 1050°C to 1200°C; however, it fell again by 0.6% as the temperature rose from 1200°C to 1300°C.

The variations in the characteristics of the coke oven liquor as a function of the wall temperature are shown in Figures 7 to 9.

The increase in temperature causes:

- a statistically significant but negligible drop in the pH value of the coke oven liquor;
- a drop in the total and free ammonia content values, with a secondary maximum corresponding to the 1200°C mark;
- a drop in the total phenol content;
- an increase in the cyanide content;
- a drop in the sulphide content as the temperature rose from 1050°C to 1100°C, followed by a rapid increase at the higher wall-temperature levels;
- the chemical oxygen requirement showed a very pronounced maximum corresponding to the 1100°C wall-temperature mark.

5. CONCLUSIONS

The test conditions adopted led to the following main conclusions:

(1) The increase in the average wall temperature from 1050°C to 1300°C caused a drop in the devolatilization time of 3.27 hours, resulting in a 25.4% increase in the specific hourly yield. The devolatilization rate increased from 27.9 to 35.0 mm/hour.

(2) The rise in the devolatilization rate resulted in an increase in coke fissuring, which was reflected in a drop of approx. 18 mm in the middle size range of the untreated coke - the average value fell from 88.6 to 70.7 mm.

The M_{40} index also fell, from an average value of 75% to one of 62%.

(3) The cold sample characteristics of the coke mechanically stabilized by crushing did not deteriorate as a result of the greater fissuring; the middle size range of the crushed coke was very stable and did not suffer from the marked variation in that of the untreated coke.

The quality indices relating to the cohesive properties of the coke (M_{10}, I_{20} and I_{10}) all improved.

(4) The mechanical properties of the coke when tested hot in a reactive atmosphere (the CSM test) improved with the rise in wall temperature up to 1200°C.

(5) The maximum in terms of quality which showed up in the hot-coke test was not repeated in the cold-sample characteristics of the cokes produced.

(6) Thermal stabilization has a positive influence above all on the mechanical hot-sample characteristics of the coke.

(7) The increase in wall temperature caused a marked variation in the pollutants present in the coke oven liquor.

Fig. 1 : CSM experimental coke oven

Table I. Chemical, plastometric and size consist characteristics of the input blends.

TYPE OF BLEND	CHEMICAL ANALISIS			BRABENDER-PLASTOGRAPH Temperature			BRABENDER-PLASTOGRAPH Viscosity			SIZE CONSIST		
	V.M. %	ASH %	SULPHUR %	Start of agglomeration °C	Maximum fluidity °C	Resolidification °C	Maximum on agglomeration Kgm.	Maximum plasticity Kgm.	Maximum resolidification Kgm.	3.36 mm %	0.15 mm %	Middle size range mm
BLEND 1	23.16	8.76	0.78	388	406	430	1.72	1.31	2.45	21.33	6.05	2.49
BLEND 2	25.24	9.34	0.80	398	407	440	1.79	1.26	2.30	26.00	5.57	2.94
BLEND 3	26.95	8.48	0.82	388	407	439	1.84	0.85	2.40	29.39	5.57	3.24

Table II. Characteristics of the devolatilization process for the various wall-temperature levels

Tempera-ture levels	TEMPERATURE		Density of wet charge	CHARACTERISTICS AT THERMAL EQUILIBRIUM		
	Average Wall Temperature	Centre of charge on pushing		Average de-volatilization time	Devolatiliz-ation rate	Specific Hour-ly Yield
	°C	°C	kg/m^3	h	mm/h	kg/m^3 h
I	1053	1016	805	16.11	27.9	46.5
II	1102	1058	808	15.91	28.3	47.0
III	1200	1121	811	13.90	32.4	53.9
IV	1306	1187	810	12.34	35.0	58.3

Table III. Results of analysis of variance.

Dependent variables	Independent variables						
	Temperature				Blend	Stabilization time	Interaction blend/temp.
	Components						
	Linear	Quadratic	Cubic	Total effect			
Middle size range of untreated coke	***			***	***		
Middle size range of crushed coke	***			***	**		
Micum M40 Index	***			***			
Micum M10 Index	***			***	**	*	
IRSID I20 Index	*	**		**	***		
IRSID I10 Index	***			**	**		
CSM hot-coke test CF2	***	***		***	**	**	
CSM hot-coke test CF1	***	***		***		**	
CSM hot-coke test CK	***	***	*	***		**	
Coke Yield $>$ 20 (untreated)	***	**		***	***		*
Coke Yield $>$ 20 (crushed)				**	***		*
pH	***	***	**	***			
Total ammonia content	**			*	*		*
Free ammonia content	*			*			
Phenol content	***	*		**	*		
Cyanide content	***			**			
Sulphide content	*			*			
Chemical oxygen requirement	***	**	***	**			

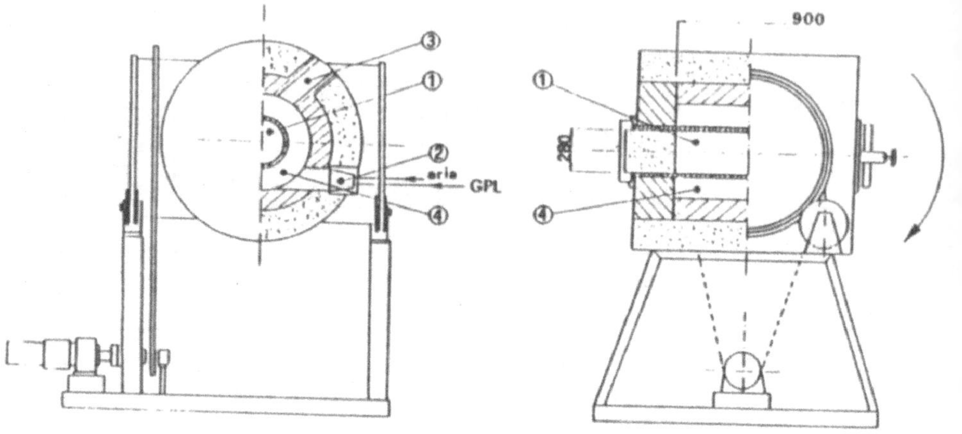

Key
(1) Test chamber
(2) LPG burners
(3) Waste gas evacuation flue
(4) Combustion chamber

Fig. 2 : CSM hot-coke test apparatus

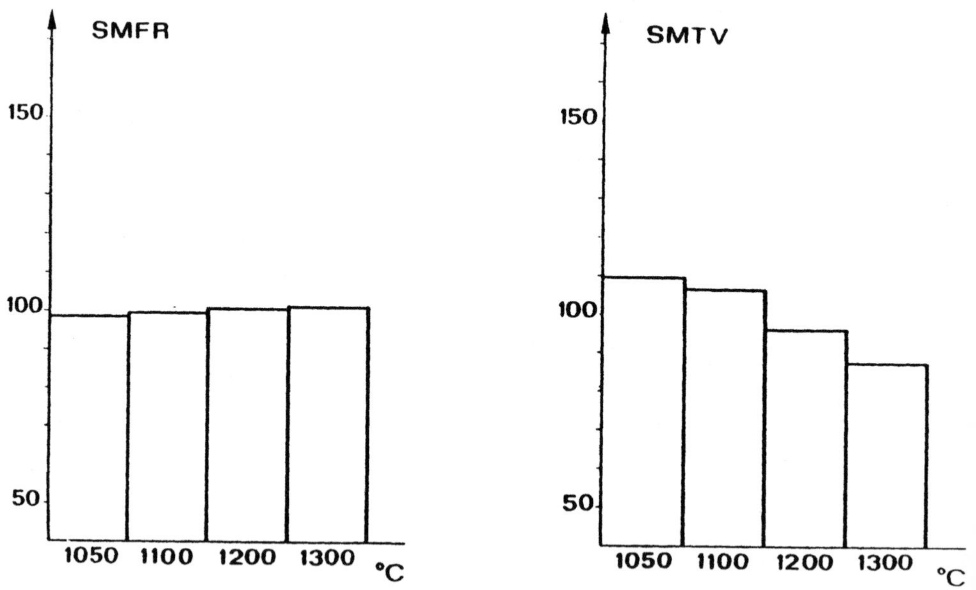

Fig. 3 : Indices for the middle size range of the untreated coke (SMTV) and the crushed coke (SMFR). (Overall average = 100)

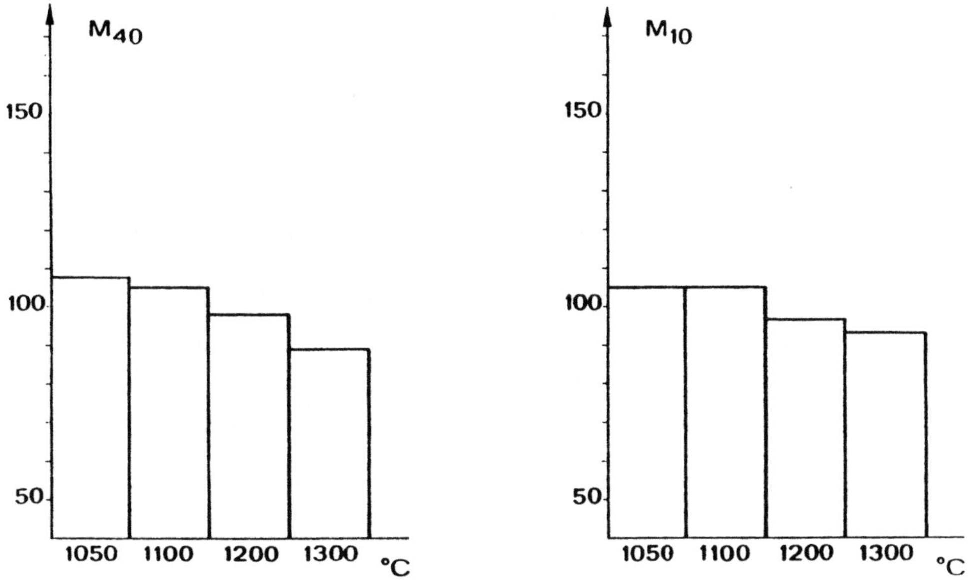

Fig. 4 : M_{40} and M_{10} indices. (Overall average = 100)

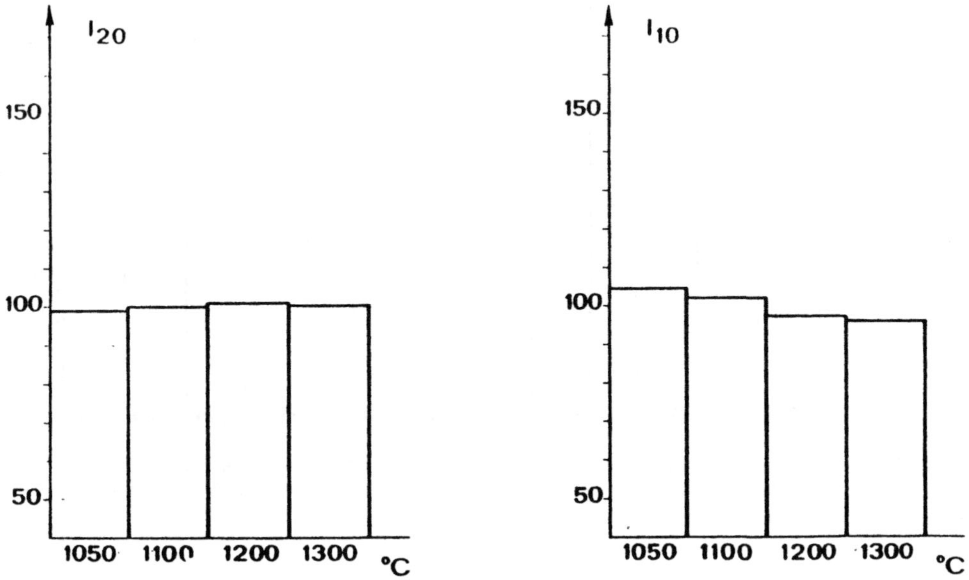

Fig. 5 : I_{20} and I_{10} indices. (Overall average = 100)

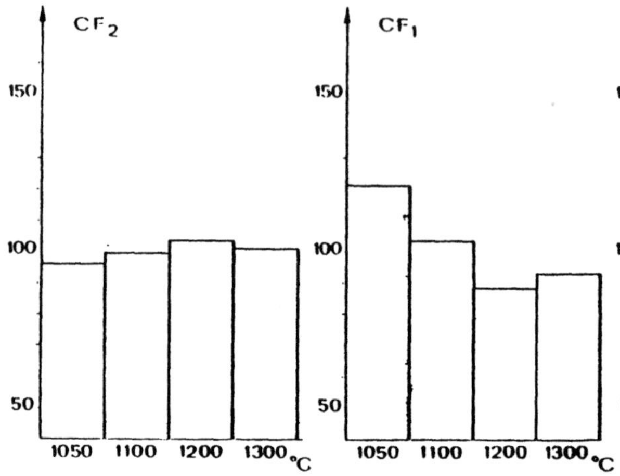

Fig. 6 : CSM hot-coke test indices. (Overall average = 100)

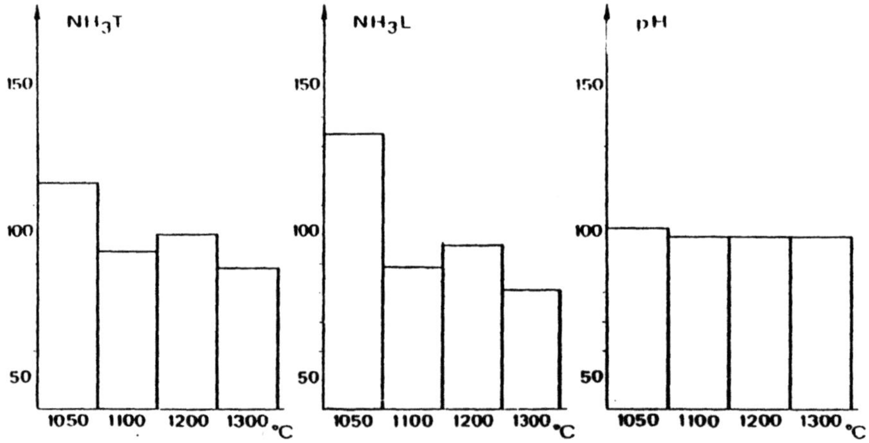

Fig. 7 : Indices for the pH value and the total ammonia and free ammonia contents in the coke oven liquor. (Overall average = 100)

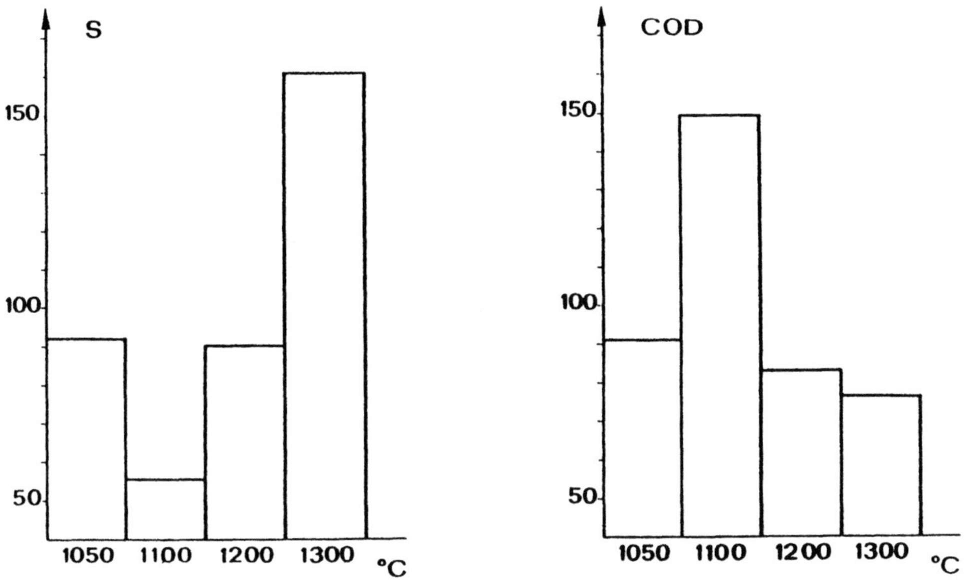

Fig. 8 : Indices for the phenol (PhOH) and cyanide (CN) content of the coke oven liquor. (Overall average = 100)

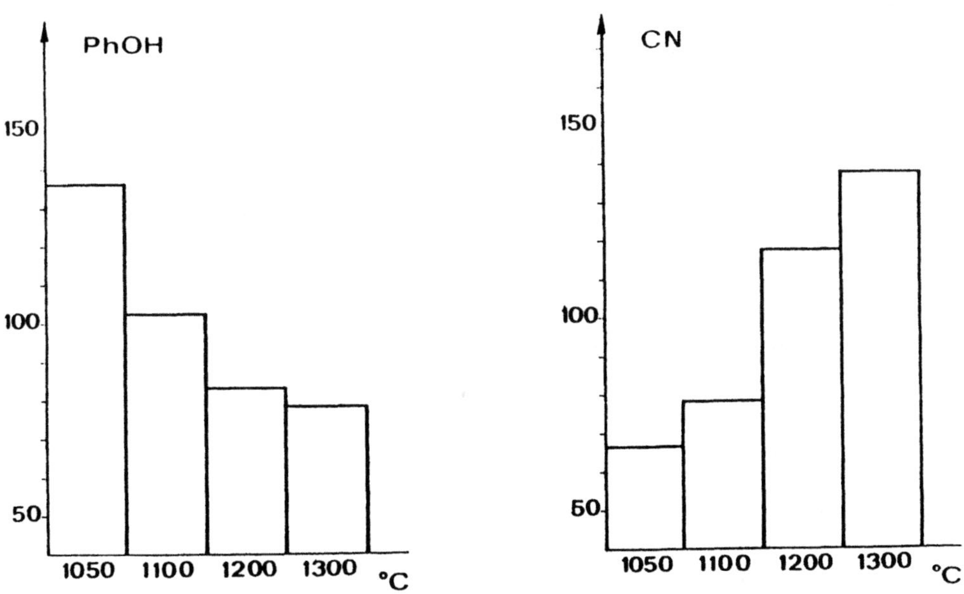

Fig. 9 : Indices for the sulphide contents and chemical oxygen requirement of the coke oven liquor. (Overall average = 100)

PARTIAL REPAIR OF DAMAGED COKE CHAMBER WALLS

N.J.W. Thijssen and P.G.J. Broersen

Research and Development Laboratory, Estel Hoogovens BV, IJmuiden

Summary

A research program was started in 1978 to find a better method for the repair of coke chamber walls.
The damage occuring after the rebuilding was thoroughly investigated as well as the repair method used. Displacement and temperature measurements were carried out during the reheating period. Large temperature gradients were found.
Bricks were tested for crack formation in a laboratory furnace, simulating the situations occuring during the repair. Critical heating rates, critical temperature gradients and maximum thermal shocks in different directions, with or without external stress, were appointed for the used silica-bricks.

A new repair method has been developed in which special attention is paid to the following points:

- the adjacent walls are kept at a temperature above 600°C
- sliding joints are applied
- expansion room for the new wall part is built in
- the supporting function of the steel construction during the reheating at the regenerator top level.

The lifetime of a repaired oven is extended and no additional damage to the adjacent walls and the heating system is detected.

1. INTRODUCTION

In 1980 Estel Hoogovens had a total coke production of nearly two million tons of coke, produced on its two coke plants. The data of both coke plants are shown in table 1.

Coke plant 1, the oldest, is equipped with four meter Coppee batteries and Coke plant 2 consists of four six and a half meter Carl Still batteries.

The investigations into the repair of damaged coke chamber walls, were carried out at Coke plant 1. However, the experience and knowledge gained during this research has given a higher priority to the maintenance of coke plant 2 in preventing coke chamber wall damage. Attention has been focussed especially on the supporting function of the steel construction and the prevention of local overheating, mainly due to irregularities in the pushing schedule.

Since 1977, as shown in figure 1, many ovens of Coke plant 1 had to be taken out of production because of failure of the coke chamber wall. Ovens were taken out of production only if they could not be charged and pushed anymore.

At that time the lifetime of the repaired ovens was mostly less than six weeks, mainly because of severe damage to the adjacent walls. This is one of the reasons for the strong increase in the number of repairs.

To overcome these and to support the discussions for a new developed repair method, more measured data were necessary.

Therefore a research project "The repair of coke chamber walls on a coke battery in operation" has been started. For this project support has been given by the ECSC. The research program has been carried out in two directions:

- Displacement and temperature measurements have been carried out during the repair and reheating of a coke chamber wall.
- In a laboratory furnace the coke oven silica-bricks have been tested individually for crack formation at temperature and pressure conditions, as found during the repair.

Based on these results a new repair method has been developed, which is used now at Coke plant 1. Until now no oven repaired with the new method has shown any damage.
However, in 1981 we have started with a complete renovation of Coke plant 1 because of the following reasons:

- The required level of coke production in the next years.
- The very low permissible emission level from the doors and the stack can not be reached by partial repairs.
- The bad overall condition of the batteries, in particular for the supporting function of the steel construction and the gastightness of the regenerators.

2. THE REPAIR METHOD USED IN 1978

The several basic elements of the repair method used at the time (figure 2) were:

- The two oven chambers next to the damaged wall are pushed empty and the gasflow to the three heating walls involved is shut-off.
- As soon as allowed by the local wall temperature, about 400°C, stopping walls are placed two flues deeper in the oven than was expected to be repaired.
- The coke chambers of both ovens behind the back stoppings are charged again and gas is put on the heating flues behind the stoppings.
- The damaged flues are demolished and new flues are built in.
- In both coke chambers stopping walls are placed near the doorframe.
- The walls are reheated by pulling holes in the back stopping walls.
- When the temperature of all three walls involved is above 800°C, the gas is switched on again, the stoppings are removed and the ovens are ready for production.

The total repair time was about three weeks in which approximately five ovens were out of production. No special precautions were taken concerning the temperature level of the adjacent walls.
An "intended" joint was used between the old and the new brickwork.
Expansion room for the new walls was left only between the doorframe and the newly built wall part for horizontal expansion.

3. TEMPERATURE AND DISPLACEMENT MEASUREMENTS DURING THE REPAIR

Temperatures were measured during the repair in the new wall, in the old wall, in the two adjacent walls and in the regenerator.
The expansion of the new wall was measured with an aid of a laser beam during the reheating.
The most important results from this measurements were:

- The measured displacements were in reasonable agreement with the calculated displacements based on the measured temperatures, except for the lower front part of the new wall.
- At the reheating the new wall expands first and after that the regenerator and thereby creating additional forces on the wall.
- The temperature of the regenerator has fallen down to about 500°C. The temperature of the end of the regenerator decreases to about 200°C.
- Large temperature differences have been found over the two adjacent walls, near the change over new - old brickwork and on the head of the new wall.

4. CRITICAL TEMPERATURE CONDITIONS FOR CRACK FORMATION IN COKE OVEN SILICA BRICKS

In our coke oven as well as for the repair, silica stretchers are used with the dimensions of 100 x 100 x 400 mm.
In a laboratory furnace these silica stretchers have been tested for crack formation under temperature and pressure conditions that could be found during the repair. From these tests it was clear that spalling of the individual stretchers could occur at the temperature conditions as found during the repair method used in 1978.

Up to the pressure level, as occurs in coke ovens (0,2 N/mm^2), no significant influence has been found on the heating conditions critical for crack formation.
To be sure that no crack formation can occur in the individual silica bricks, because of the internal stresses caused by the temperature differences over the brick, the following practical rules have been taken:

- The critical heating rate for the coke oven silica stretchers is two to three times lower when the heat flows through the 400 x 100 mm surface than through the 100 x 100 mm surface.
  To be sure that in the individual bricks no crack formation occurs because of the heating rate, this heating rate must be lower than 50°C/h.
- At a temperature level of 200°C the linear temperature gradient should not exceed 1000°C/m.
- The temperature of the connecting surface between the old and new bricks should not exceed 140°C at the bricklaying.
  This means that in case of a direct contact (no isolation layer) between the new and old bricks, the maximum allowable temperature of the old bricks is then 250°C.

5. DISCUSSION ON THE FAILURES OF THE OLD REPAIR METHOD

Based on the temperature and displacement measurements, model calculations of the oven wall, tests on the individual bricks and the inspection of several repaired oven walls, we have explained the observations as described below:

- A severe crack has been detected along the headers near the change over of the old and new brickwork (figure 3).
  This crack is so severe because the not straight joint between the old and new brickwork has no room to meet the expansion of the regenerator.
- A crack in the new wall along the header near the front stopping is detected. The measured temperature difference over these bricks during reheating was 200°C over 10 cm at a temperature level of about 200°C. From the results obtained in the laboratory furnace it is clear that this can cause cracks.

- The open crack in the regeneratorwall (figure 4) and the downward movement of the lower section of the new built wall near the doorframe.
  We explain this as follows:
  We build a new wall on a cold regenerator that has not shrinked totaly in the horizontal direction. Now first the new wall is expanding and hereby opens the crack in the regenerator when this one is not enough supported by the steelconstruction.
- The rapid decay of the adjacent old walls.
  From the measured temperature differences over these walls during the repair and from the results of our test oven, it is clear that a lot of small cracks develop during the repair in the old walls. We believe that this is the main reason for the rapid wear of these walls.
- When the oven is taken in production again, the oven has often heavy pushers or stickers because of the bending of the walls. The main cause for this is that no vertical expansion room was built in between the new wall and the ovendeck.

6. THE NEWLY DEVELOPED REPAIR METHOD

The basic points for the new repair method are no linear temperature differences bigger than 1000°C/m and built-in freedom for the thermal expansion of the bricks. This means the following changes with respect to the old repair method:

- Connection old - new brickwork

  Here we use now a vertical expansion joint of two millimeters. The joint is filled up with isolation fibre. The new wall can freely expand in vertical direction without adding additional stresses to the bricks. By using the isolation fibre the temperature of the old bricks can stay up to 250°C without damaging the new bricks applied to it.

- Horizontal expansion joint

  At the (upper) layer of the newly built wall a horizontal expansion joint is left of 60 mm filled with isolation fibre. In this way the bending of the new wall is prevented.

- Head end of new wall

  For the fist brick a semi-silica quality is used and a vertical expansion joint is used for the connection with the rest of the wall.

- After reheating all the open joints get a spraying treatment to close them before the oven is put in production again.

- Location of the stopping walls

  The back stopping wall must be placed two flues deeper in the oven than necessary for the rebuilding. The front stopping must be located as far as possible to the front to prevent the new wall from direct contact with the outside temperatures.

- The adjacent walls

  Both the adjacent walls have to be isolated before the rebuilding starts. The temperature on the burner side has to stay always above 600°C to prevent crack formation.

  Because of the construction of our Coppee-batteries, this has also the advantage that the regenerator stays on a higher temperature. By this the expansion of the regenerator after repair is less and therefore the risk of formation of additional open joints is decreased.

- The supporting steelconstruction

  Without a good supporting steelconstruction no repair gives on a longer run a good result. Special attention must be given to the support of the upper layers of the regenerator.

7. CONCLUSION

The basic reasons for the failure of our coke chamber walls are:

- The steelconstruction did not always give the necessary support to the brickwork.
- Irregularity in the realised pushing schedules.

By this open joints are created, this causes local overheating, that is followed by extreme wear of the bricks by melting coal ash or spalling and at the end collapse of the coke chamber wall.
The method used to repair this damage has created additional damage to the adjacent walls and reheating system.
Our research project on this subject has quantified several aspects and has been the basis for development of a new repair method which does not create additional damage to the adjacent walls or heating system.
The main elements of this new method are:

- Keep the individual bricks within the critical temperature range.
- Build in room for the expansion of the walls.
- Cool down the old brick work as little as possible.
- Take care of the supporting function of the steelconstruction during the reheating, especially around the ovensole.

The whole repair must be carried out very accurately because it is easy to create additional cracks, but very difficult to repair them.

The repair of coke chamber walls is expensive. In our opinion more attention and money than usual have to be spent to the maintenance of the supporting function of the steelwork, the gastightness of the brickwork and to the regularity of the pushing schedule.

TABLE 1: Oven data.

Battery no.	Type	Start up	Number of ovens	Width of ovens (mm)	Height of ovens (m)	Length of ovens (m)	Average coking time (1980) (hrs)
11	Coppee	'68	36	460	4.0	11.7	25.8
12	Coppee	'67	36	460	4.0	11.7	25.8
15	Coppee	'55	30	460	4.0	11.7	26.8
16	Coppee	'55	30	460	4.0	11.7	26.8
17	Coppee	'59	36	460	4.0	11.7	26.8
18	Coppee	'63	36	460	4.0	11.7	26.8
21	Still	'72	27	420	6.5	15.5	19.4
22	Still	'72	27	420	6.5	15.5	19.4
23	Still	'72	27	420	6.5	15.5	19.3
24	Still	'72	27	420	6.5	15.5	19.5

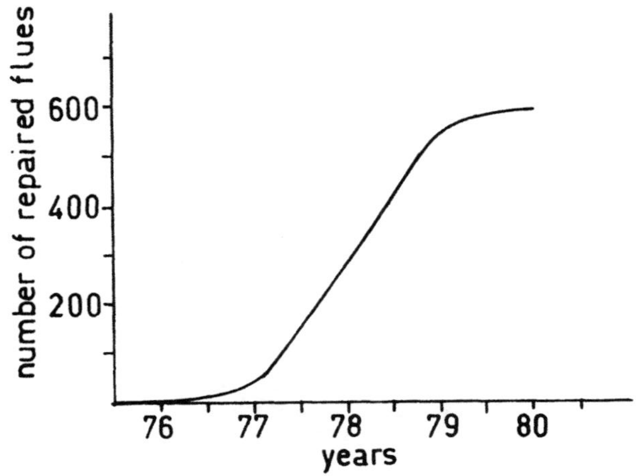

FIGURE 1: Number of flue repairs in the last 5 years.

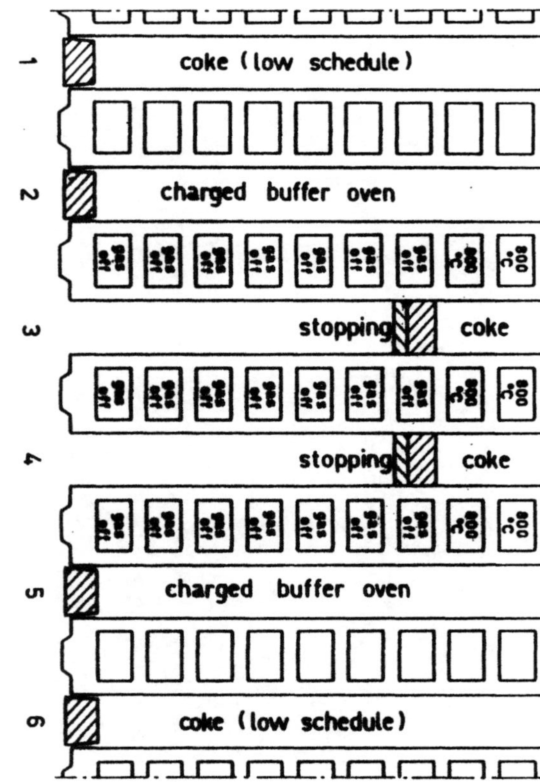

FIGURE 2: Cross section of a battery part during repair.

FIGURE 3: Crack at the intended joint between the old and the new brickwork.

FIGURE 4: Open crack at the end of the regeneratorwall after a repair according to the old method.

ROUND TABLE "A"

COKE AND THE BLAST FURNACE

Chairman: G. SIRONI

The following conclusions, to be verified at future meetings, may be drawn at the end of this highly interesting symposium on coking plant techniques, which focused on coke and blast furnaces.

- The mechanical properties required of coke at ambient temperature are a necessary condition in many cases for achieving virtually trouble-free blast furnace operation, but they are not sufficient to ensure high performances and consistent yields of metal.

- The properties of coke when hot must also be taken into account.

- During the round-table discussion held yesterday, these two aspects received most attention.

- The property of hot coke uppermost in speakers' minds was "reactivity" i.e. the propensity of coke in particular to react with carbon dioxide (solution loss).

- It should be made clear at the outset that a certain degree of solution loss or direct reduction (i.e. reduction involving direct consumption of carbon) is necessary to keep the thermo-chemical conditions of the blast furnace process stable around the minimum fuel consumption level.

- Like mechanical resistance, reactivity is related to the coke's physical macrostructure, and studies to determine the porosity and specific surface area of coke are particularly significant. Correlating these with coal blends and coking conditions will enable production technology to harness such knowledge.

- Of equally fundamental importance is research into the formation and properties of the microstructure and the optical texture formed during the coking process.

The optical texture proves to be related to the properties of coke for industrial use and more particularly to its behaviour during gasification and its resistance to alkalis.
Research in this field is of value because it enables the behaviour of coke to be explained, and so rationalizes and gradually reduces the number of tests required to determine the technological behaviour of coke.

The problem of alkalis in blast furnaces and their effects on the coke may be summarized as follows.
It is a task of the furnace operator to reduce the alkalis present in blast furnaces by:

a. Selecting the materials in the charge;
b. Avoiding the build-up of alkalis in the furnace by using slag to help eliminate them. This may be achieved by altering the composition of the slag, at least from time to time, in terms of the $\frac{CaO+MgO}{Si\ O_2}$ ratio, i.e. by reducing this ratio with respect to normal operating conditions.

- As emerged from the discussion and the information given in the introductory paper, variations in coke reactivity within the range encountered in good coking plant conditions lead to differences of between 1 and 2% in specific consumption of coke in blast furnaces.

- Less widely known but perhaps of greater importance are the effects of the structural weakening of coke in blast furnaces caused by the loss of carbon due to solution loss.

- To this should be added the effect of the rapid heating of the coke to temperatures above that needed for coking. (The effect of high temperatures on the stability of coke is the special field of certain research centres in Europe)

- The combined chemical and thermal effects are probably the cause of coke size degradation in blast furnaces.

- This substantial degradation in coke particle-size was shown by studies carried out in Japan on several blast furnaces allowed to cool down at the end of a run. There was a decrease of up to 50% in average particle size.

- This particle-size degradation already taking place in blast furnace boshes is perhaps the most serious consequence of (structurally) excessive coke reactivity, the effect of alkalis and probably the high temperatures.

- The coke particle-size degradation is accompanied by a deterioration in permeability in the bosh, the most critical zone, and by slower, less smooth running of the blast furnace, which also deteriorates from the point of view of the quality of the metal produced, with effects on the intervals between and the length of blast furnace runs, or in other words on the useful life of the plant.

- While stress should be laid on the major contributions presented at the symposium on subjects related to the various aspects of the behaviour of coke, the continuation of such research seems highly important on account of the impact it may have in the near future on knowledge of an operating procedure for coking plants and blast furnaces.

In spite of the fact that interest during the round table centred on coke reactivity (i.e. the way coke behaves in blast furnaces), other subjects of interest to the steel industry which were dealt with during the two days should not be neglected.

1. Techniques aimed at widening the range of coals used in coking or improving quality using specific blends. Various methods worth mentioning are: preheating the coal (Precarbon), stamping or compacting the charge partial briquetting of the charge and coals of lower coking quality (Bricoke).

In addition, there is a certain amount of interest in formed, coke pending wider industrial developments. It should be borne in mind that the European Economic Community imports almost 30% of its coking coal, and in Italy, to quote on example, imports account for the total.
In these circumstances any endeavour to make use of a wider range of possibly cheaper coals takes on considerable economic importance.

Lastly, mention should also be made of research into coking rates and the stabilization of coke on account of the practical benefits they have for the running of major steel industry coking plants, which at times need to adapt their operations to varying demands, obviously over long periods.

Such adjustments may be quantitative or they may be stricter requirements regarding the technological properties of the coke.

- The now urgent need for an improvement in the level of technical control of the batteries is stressed in a number of papers, and innovation in plant is dealt with in the paper presented by CODECO.

In conclusion, this symposium gave a clear picture of the state of research and technological innovation in the EEC. Everything pointed to the fact that the guidelines for both scientific and technological research were well chosen, and there need be no fear of failure to produce the results which are vital for the constant modernization in Europe's coking and steel industries.

ROUND TABLE "B"

POLLUTION PROBLEMS

Chairman: J.P. GRAHAM

Mr Chairman, Ladies and Gentlemen,

This session differed from others in the Symposium in that there was only one relevant technical paper, a contribution from Mr R. F. Littlejohn and Mr M. A. Wright of the National Coal Board entitled "Emission of tar fog from coke-oven doors". In his presentation, Mr Littlejohn referred to the levels of BSM arising from door leakage which he highlighted as the main contributor of BSM emissions. He also confirmed the relationship between visual, subjective, leakage monitoring and active BSM emissions but suggested that the actual relationship is not the same for each plant, although further work was presently being carried out to investigate this aspect of the studies. The discussion of this paper led to comments on oven door design and, in particular, reference was made to current work in this area supported by ECSC funds. Dr Eisenhut mentioned the co-operation between the UK and the Federal Republic of Germany in studies of the assessment and quantification of leakage prior to the programme of studies on causes of door leakage and means of mitigation. He referred to interesting developments of doors with double seals and the effectiveness of flexible seals to minimize door leakage - a 65 per cent reduction being possible. The significance of poor maintenance and lack of effective mechanical cleaners were cited as significant causes of door leakage.

Mr Kirk commented on the importance of the conclusions in Mr Littlejohn's paper in that the BCRA visual assessment technique originally designed to reflect standards of operation and maintenance also reflected actual changes in BSM emissions. Mention was also made in this context of the forthcoming HSE document which required standards of leakage control based on the BCRA uniform method.

The fact that there was only one paper prompted the expression of a view that this in some way reflected the great progress which had been made in pollution control in the past decade in such areas as the nuisance caused by grit and dust emissions.

This view was supported by reference to the practical steps which had been taken on some older batteries where preventitive measures had been taken with some success. In this context, it was suggested that much progress had been made at relatively small cost, a point noted by the Chairman, who referred to the emphasis placed on the importance of good housekeeping at coke ovens by the UK Alkali and Clean Air Inspectorate.

A contrary view was put forward which expressed the views of the pollution control authorities in Germany to the effect that whilst some improvements were acknowledged in a variety of circumstances on coking plants, there were nevertheless areas in which available means of pollution control had not been applied. There also remained the problem of fumes and sulphur gases which were still not contained.

There was evidence that whereas in the past there had been considerable pressure on the coking industry to mitigate the nuisance caused to areas in the immediate vicinity of coking plants, much greater emphasis was now being attached to worker protection - witness the proposed legislation in Germany in connection with threshold limit values. Additionally, there was the predicted legislation in the UK in regard to recommendations for best operational practice. The meeting expressed some concern that the TLVs now being suggested by the authorities could not be achieved any way, despite extremely heavy capital expenditure and running costs associated with the equipment. This was a matter of very serious concern in the indsutry, particularly in times of recession.

Whilst much of the discussion had ranged upon the aleviation of the consequences of coal carbonization in coke ovens, the view was expressed that attention should more properly be devoted to the causes of pollution. This was based on the view that by building larger oven chambers, less ovens would be required for a given coke output and therefore the number of potential sources would be reduced. Support for this concept should be the concern of the Community, as should be the realization that some of the proposed legislation would be setting technically unachievable targets. It was however recognized that the smaller number of sources which the larger oven chambers implied would not necessarily be easier to deal with as the increase in oven door dimensions, for example, might in itself present problems. The main points expressed during the Round Table may therefore be summarized as:

(a) We have made a great deal of progress in recent years in connection with the mitigation of atmospheric pollution from coke oven plants.

(b) The change in emphasis is towards worker protection and the improvement in the atmosphere in which the operators work rather than the mitigation of nuisance outside the works boundary, the latter having been substantially corrected.

(c) The new threshold limit values now being proposed in some legislation would be difficult, if not impossible, to achieve.

(d) At the present time, the industry can ill afford the very high capital costs - possibly as much as 20 per cent - for the establishment of environmental control, particularly bearing in mind the technical inability of achieving some of the TLVs now being suggested.

CLOSING ADDRESS

A. DE GREEF

Ladies and Gentlemen,

After having spent two days in each others' company, this symposium is now coming to a close with some of us still thirsting after knowledge and others totally satiated. I hope you will not mind my adding some closing remarks.

Firstly I should like to convey my own and the Commission's thanks to all the people who, in whatever capacity, either in the limelight, backstage or at organization level, have helped to make a success of this symposium. Without them, and particularly without their close collaboration, such gatherings would be impossible to organise and we would be unable to gain such a wide audience for the interesting and up-to-date information which you have now been given in a research field as significant as that of coking plant techniques.

This field is one of the major focal points of coal beneficiation, since coke is still a fruitful outlet for the Community coal industry and is a fundamental part of the manufacture of iron and steel products.

The past two days, have given us the first opportunity since 1975 to update our knowledge of the progress made, in a pleasant atmosphere and with total mutual comprehension.

In addition to learning something from the symposium, we may perhaps have met old acquaintances or school friends, or struck up new friendships with people for whom we were previously no more than names.

We have made every effort to provide you with as much information as possible in the booklet we have published. We shall follow this up later by sending you the minutes of the technical meetings and panel discussions as soon as we are able.

I feel sure that the research work carried out in this field, with Community-wide cooperation, has not only furthered the cause of science, but also helped towards accomplishing the at times difficult task of building tomorrow's Europe. The success of the past two days is real proof that this is being achieved.

I sincerely thank all those who have taken part in this symposium and those who have contributed to the discussions.

I am also grateful to:

Mr Reichert for having opened this symposium;
the chairmen of the technical meetings and panel discussions;
the speakers;
the Technical and Scientific Preparation Committee;
the Directorate-General for Information Market and Innovation and the Directorate-General for Personnel and Administration, particularly the Luxemburg departments, and the staff and interpreters of the Joint Service Interpretation - Conferences;
Mr Linster and Miss Goebel, not forgetting Mr Rotondo';
and lastly, Mr Wilkinson and Miss Gassen who work with me.

With these words of thanks I hereby declare this symposium closed and wish you all a safe journey and return home.

LIST OF PARTICIPANTS

ABEL, O.	Institut für Chemische Technologie und Brennstofftechnik Technische Universität Clausthal Erzstrasse 18 D - 3392 CLAUSTHAL ZELLERFELD 1
ALESSI, W.	ITALIANA COKE S.p.A. Via Giovannino de Grassi, 8 I - 20123 MILANO
ATES, F.	ENSCH S.à.r.l. P.O. Box 2132 L - 1021 LUXEMBOURG
AULAGNIER, S.	CERCHAR 33, rue de la Baume F - 75008 PARIS
BAMBERG, E.H.	DEUTSCHE SHELL AG Hohe-Schaar-str. 36 D - 2102 HAMBURG A3
BARBER, K.G.	OTTO SIMON CARVES P.O. Box 57 GB - STOCKPORT SK3 OTU
BARBIER, C.	S O L M E R F - 13776 FOS SUR MER Cedex
BAUER, H.	R U H R K O H L E AG Rüttenscheider Strasse 1 D - 4300 ESSEN 1
BAZIN, P.	SHELL FRANCAISE 29 Rue de Berri F - 75397 PARIS Cedex 08
BECK, K. G.	BERGBAU-FORSCHUNG Franz-Fischer-Weg 61 D - 4300 ESSEN 13
BECKER, W.	BERGBAU AG LIPPE Shamrockring 1 D - 4690 HERNE 1

BECKMANN, F.	RUHRKOHLE AG Postfach 1o3262 D - 4300 ESSEN 1
BECKMANN, F.H.	Direktor ESCHWEILER BERGWERKS-VEREIN AG Postfach 1345 D - 5110 ALSDORF
BECKMANN, R.	MANNESMANNROEHREN-WERKE AG Betr. Abt. Hüttenwerk D - 4100 DUISBURG 25
BERNARD, J.	Directeur HOUILLERES DE LORRAINE - HBL 2, rue de Metz F - 57800 FREYMING-MERLEBACH
BERNET, H.G.	S O F R E S I D 59, rue de la République F - 93108 MONTREUIL/BOIS
BERTLING, H.	RUHRKOHLE AG Postfach 103262 D - 4300 ESSEN 1
BIEHL, K.	INTERCOKE Rue Bausbeau L - LUXEMBOURG
BIESHEUVEL, A.	STAALBEDRIJF IJMUIDEN ESTEL HOOGEVENS BV Postbus 1o.ooo NL - 1970 CA IJMUIDEN
BILLOT, F.	S.A. BELREF Rue de la Rivierette 1oo B - 7330 SAINT GHISLAIN
BINEK, H.	H B N P C Boîte Postale 513 64, rue des Minimes F - 59505 DOUAI, Cedex
BLASE, M.	DIDIER ENGINEERING GmbH Alfredstrasse 28 D - 4300 ESSEN 1

BOECKER, D.	RHEINBRAUN AG Stüttgenweg 2 D - 5000 KOELN 41
BONAVENTURE, M.	COCKERILL - SAMBRE S.A. B - 6090 CHARLEROI (Couillet)
BOTTOIFSEN, A.E.	Managing Director NORSK KOKSVERK A/S P.O. Box 2o3 N - 8601 MOIRANA
BOWNESS, M.M.	BRITISH STEEL CORPORATION Redcar Works GB - REDCAR
BRACCO, T.	ITALIANA COKE S.p.A. Via Giovannino de Grassi, 8 I - 20123 MILANO
BROERSEN, P.	Hoogevens Estel BV Centraal Laboratorium 2 - 3G.16 Postbus 1o.ooo NL - 1970 CA IJMUIDEN
BROOKS, S.H.	Director BRITISH STEEL CORPORATION c/o Whithead Works P.O. Box 28 GB - NEWPORT, GWENT NPT 2WX
BROWN, K.	Director CARBON CONSULTING INT. 22, Lambton Road BROADMEADOW, N.S.W. 2292 AUSTRALIA
BUECKEN, H.	Direktor AKTIEN-GESELLSCHAFT DER DILLINGER HUETTENWERKE Postfach 1580 D - 6638 DILLINGEN/SAAR
BUSSMANN, B.	THYSSEN AG Kaiser Wilhelm Strasse 1oo D - 4100 DUISBURG 11

BUYSSCHAERT, J.P.	CDF INGENIERIE 2 route de la Bassée F - 62300 LENS
CHARPENTIER, G.	Cokerie de PONT-A-MOUSSON SA F - 54700 PONT-A-MOUSSON
COOPER, D.A.	ISCOR - HEADQUARTERS PO Box 450 SOUTH AFRICA - PRETORIA 0001
CRELLING, J.C.	Department of Geology Southern Illinois University USA CARBONDALE ILLINOIS 62901
DE ARIZON, L.	T R A C T I O N E L Rue le la Science 31 B - 1040 BRUXELLES
DE BACKER, J.R.G.	CORPS DES MINES Rue de Corbais 67 B - 5873 HEVILLERS
DECKERS, F.	Divisiedirekteur der Mijner MIJNWEZEN - EK. ZAKEN Demerstraat 81 B - 3500 HASSELT
DE DEMO, E.	Administrateur-Directeur IMC - LUXEMBOURG S.A. 3 Place Winston Churchill L - 1340 LUXEMBOURG
DEFLANDRE, H.	Directeur Studiedienst N.V. KEMPENSE STEENKOLENMIJNEN Grote Baan 27 B - 3530 HOUTHALEN-HELCHTEREN
DE FRANCO, F.	ITALSIDER I - 74100 TARANTO
DE GOUBERVILLE, M.	KOEG, KOHLENEINKAUFSGES Feldmannstrasse 113 D - 6600 SAARBRUECKEN

DE GREFF, A.Y.J.	Commission des Communautés européennes Chef de division Diréction générale "Energie" Rue de la Loi 2oo B - 1049 BRUXELLES
DELANNAY, B.	HOUILLERES BASSIN LORRAINE 2, rue de Metz F - 57802 FREYMING-MERLEBACH
DELANNAY, G.	H.B.N.P.C. Boîte Postale 513 64, rue des Minimes F - 59505 DOUAI Cedex
DELANNOY, G.	Directeur général Adjoint Cdf - HBCM 9, avenue Percier F - 75360 PARIS Cedex 08
DELESSARD, S.	C E R C H A R Station Expérimentale de Marienau F - 57600 FORBACH
DELLA CASA, H.	Directeur C E S S I D Boîte Postale 56 F - 57210 MAIZIERES LES METZ
DEMAREZ, A.	S.A. CARCOKE Rue de Chièvres B - 7340 TERTRE
DEPREZ, J.	S.A. HAINAUT - SAMBRE Cokerie de Marchienne Rue de la Docherie B - 6030 MARCHIENNE AU PONT
DERYCKE, J.	S.A. SIDMAR Pres.J.F. Kennedylaan 51 B - 9020 GENT
DESWAEF, R.	INIEX Rue du Chèra 2oo B - 4000 LIEGE

de TORCY, B.	C E C ENTREPRISE 113 rue Jean Marin Naudin F - 92220 BAGNEUX
de WOOT de TRIXHE, P.	S.A. COCKERILL - SAMBRE B - 4100 SERAING
DITTE, F.	CDF INGENIERIE 2 route de la Bassée F - 62300 LENS
DONCKEL, G.	I N T E R C O K E 4 rue Bourbon L - 1249 LUXEMBOURG
DUBASH, R.	Works Manager NATIONAL SMOKELESS FUELS LTD. Derby Road GB - CHESTERFIELD DERBYSHIRE S42 6JT
DUCHENE, J.M.	C E R C H A R Station Expérimentale de Marienau Usine de Marienau F - 57600 FORBACH
DUNGS, H.	FIRMA CAR STILL Postfach 1480 D - 4350 RECKLINGHAUSEN
DURSELEN,	KRUPP-KOPPERS GmbH Postfach 1o2251 D - 4300 ESSEN 1
DUTTA-ROY, S.	DIDIER ENGINEERING GmbH Alfredstrasse 28 D - 4300 ESSEN 1
ECHTERHOFF, J.	KOKEREIGESELLSCHAFT SAAR mbH Feldmannstrasse 113 D - 6600 SAARBRUECKEN
EISENHUT, W.	STEINKOHLENBERGBAUVEREIN Postfach 13o14o D - 4300 ESSEN 13

EISMANN, D.E.	DIDIER ENGINEERING GmbH Alfredstrasse 28 D - 4300 ESSEN
ENGLISCH, N.	BERGBAU AG LIPPE Shamrockring 1 D - 4690 HERNE 1
EPPELS, K.	RUHRKOHLE AG Postfach 1o3262 D - 4300 ESSEN 1
ESCOIN, M.	HOUILLERES DE LORRAINE Cokerie de Corling F - 57802 FREYMING-MERLEBACH
FALK, E.	RUHRKOHLE AG Postfach 103262 D - 4300 ESSEN 1
FEUGERE, J.	SOCIETE PENNAROYA F - 62950 NOYELLES-GODAULT
FIEDLER, J.	LANDESOBERBERGAMT NW Goebenstrasse 25 D - 4600 DORTMUND 1
FIORINI, E.	ITALIANA COKE S.p.A. Via Giovannino de Grassi 8 I - 20123 MILANO
FIRKET, J.	Directeur S.A. CARCOKE 216 Chaussée de Vilvorde B - 1120 BRUXELLES
FISCHER,	KRUPP-KOPPERS GmbH Postfach 1o2251 D - 4300 ESSEN 1
FLASCHE, K.H.	STAHLWERKE-ROCHLING-BURBACH D - 6620 VOELKLINGEN

FLOCKENHAUS, C.	DIDIER ENGINEERING GmbH Alfred Strasse 28 D - 4300 ESSEN 1
FONTEYN, A.	ADMINISTRATION DES MINES Rue Jos de Swerts 58A B - 2060 MERKSEM
FRANCK, A.	HOUILLERES DU BASSIN DE LORRAINE Cokerie de Marienau F - 57600 FORBACH
FRANK, K.	GREAT LAHER CARBON Park Avenue 299 U.S.A - NEW YORK
FRENAY, C.	DIVISION DU HAINAUT ADMINISTRATION DES MINES Centre Albert Place Albert 1er B5 - II étage B - 6000 CHARLEROI
FRICK, H.	KOKEREIGESELLSCHAFT SAAR mbH Feldmannstrasse 113 D - 6600 SAARBRUECKEN
FRIEBE, G.	MANNESMANNROEHREN-WERKE AG Betr. Abt. Hüttenwerk D - 4100 DUISBURG 25
GAILLET, J.P.	C E R C H A R Station Expérimentale de Marienau F - 57600 FORBACH
GALOW, M.	DIDIER ENGINEERING GmbH Alfredstrasse 28 D - 4300 ESSEN 1
GANIER, M.E.	C E R C H A R 72, Rond Point du Pont de Sèvres F - 92100 BOULOGNE
GARCIA-MORAN, J.M.	INDUSTRIAL QUIMICA NALON S.A. P.O. Box 153 E - OVIEDO

GASSEN, C.	KOMISSION DER EG Rue de la Loi 2oo B - 1049 BRUXELLES
GERVASONI, M.	ASSIDER 8 Piazza Velasca I - 20122 MILANO
GHIGNY, P.	S.A. FORGES DE CLABECQ Div. Fours à Coke Slachthuisstraat 138 B - 1800 VILVORDEQ
GHYSELS, P.	KEMPENSE STEENKOLENMJNEN N.V. Grote Baan 27 B - 3530 HOUTHALEN-HELCHTEREN
GOEDDE, F.	Kokereidirektor EBV HERZOGENRATH Roermonderstrasse 63 D - 5120 HERZOGENRATH-KOHLSCHEID
GOLECZKA, J.	NATIONAL COAL BOARD, C.R.E. Stoke Orchard GB - CHELTENHAM, GL52 4RZ
GRAAS, A.	SACILOR - SOLLAC S.C.E. Boîte Postale 11 F - 57190 FLORANGE
GRAHAM, J.P.	THE BRITISH CARBONISATION RESEARCH ASSOCIATION Wingerworth, Chesterfield GB - DERBYSHIRE S42 6JS
GRINA, J.A.	U.S. STEEEL CORPORATION 6oo Grant Street U.S.A. - PITTSBURGH, P.A. 15230
GROSS, H.	Betriebsdirektor SAARBERGWERKE AG Kok.Fürstenhaüsen D - 6620 VOLKLINGEN-FURSTENHAUESEN

GROSS, R.
DILLINGER HUETTE AG
Postfach 158
D - 6638 DILLINGEN / SAAR

GUERIN, R.
CDF Internationale
BP 396 - 08
F - 75360 PARIS Cedex 08

HABERMEHL, D.
BERGBAU-FORSCHUNG
Franz-Fischer-Weg 61
D - 4300 ESSEN

HAKE, P.
RWTH AACHEN
Josef Ponten Str. 77
D - 5100 AACHEN

HANSEN, P.
ADMINISTRATION DE L'ENVIRONMENT
1a, rue Auguste Lumière
L - 1950 LUXEMBOURG

HARDT, W.
SAARBERGWERKE AG
Postfach 1o3o
D - 6600 SAARBRUECKEN

HARION, E.
SAARBERG-INTERPLAN GmbH
Stengelstrasse 1
D - 6600 SAARBRUECKEN

HARVENG, I.
HAINAUT - SAMBRE
Rue de l'Usine
B - 6090 COUILLET

HAYWOOD, S.P.
OTTO - SIMON CARVES
P.O. Box 57
GB - STOCKPORT

HENNIGFELD, H.
EBV AG
Robert-Koch-Strasse
D - 5110 ALSDORF

HERMANN, W.
ESCHWEILER BERGWERKS-VEREIN A.G.
Roermonderstrasse 63
D - 5120 HERZOGENRATH-KOHLSCHEID

HOFHERR, K.	Betriebsdirektor THYSSEN AG Postfach 11oo67 D - 4100 DUISBURG 11
HOLZ, T.	ESCHWEILER BERGWERKS-VEREIN AG Postfach 1345 D - 5110 ALSDORF
HOERSKEN, D.	Betriebsdirektor BERGBAU AG LIPPE Shamrockring 1 D - 4690 HERNE 1
HOSTI, L.	U S I N O R F - 54230 NEUVES-MAISONS
HURST, H.	BRITISH STEEL CORPORATION BSC REDCAR WORKS GB - REDCAR
IGUCHI, T.	NIPPON STEEL CORPORATION Königsallee 3o D - 4000 DUESSELDORF
INESON, A.	Production Director NATIONAL SMOKELESS FUELS LTD Eastwood Hall GB - EASTWOOD, NOTTINGHAM AG16 3EB
IVANIER, L.	I R S I D F - 57210 MAIZIERES LES METZ
JOKISCH,	KRUPP-KOPPERS GmbH Postfach 1o2251 D - 4300 ESSEN 1
JORZYK, S.	SAARBERGWERKE AG Trierer Strasse 1 D - 6600 SAARBRUECKEN
JOSEPH, H.	DIDIER ENGINEERING GmbH Alfredstrasse 28 D - 4300 ESSEN 1

JURANEK, G.	RUHRKOHLE AG Postfach 1o3262 D - 4300 ESSEN 1
KARBOVIAC, D.	C E R C H A R Station Expérimentale de Marienau F - 57600 FORBACH
KARTE, W.	STAALBEDRIJF IJMUIDEN ESTEL HOOGEVENS BV Postbus 1o.ooo NL - 1970 CA IJMUIDEN
KASPERCZYK, J.	ENSCH S.à.r.l. Boîte Postale 2132 L - 1021 LUXEMBOURG
KERSTING, H.J.	RHEINBRAUN AG Stüttgenweg 2 D - 5000 KOELN 41
KIRILINE, V.	REPRESENTATION COMMERCIALE DE L'URSS Bd. du Régent 31-33 B - 1000 BRUXELLES
KIRK, B.P.	THE BRITISH CARBONISATION RESEARCH ASSOCIATION Wingerworth, Chesterfield GB - DERBYSHIRE S42 6JS
KIRSH, G.	ESCHWEILER BERGWERKS-VEREIN AG Postfach 1345 D - 5110 ALSDORF
KLEIN, F.	C.E.C.E.S. Boîte Postale 36 Voie Romaine F - 57210 MAIZIERES LES METZ
KNOCHE, G.	DIDIER ENGINEERING Alfredstrasse 28 D - 4300 ESSEN 1

KOVPAK, V.	REPRESENTATION COMMERCIALE DE L'URSS Bl. du Régent 31-33 B -1000 BRUXELLES
KROES, M.	STAALBEDRIJF IJMUIDEN ESTEL HOOGEVENS BV Postbus 1o.ooo NL - 1970 CA IJMUIDEN
KURTZ, R.	RHEIN. BRAUNKOHLENWERKE AG Postfach 89o3 D - 5030 HUERTH
KUTZNER, R.	RUHRKOHLE AG Postfach 1o3262 D - 4300 ESSEN 1
LAHAYE, J.	Directeur Technique S. B. M. Parc de Recherche de Sant Tilman B - 4900 ANGLEUR PAYS
LANGENBECK, W.	KRUPP STAHL AG Postfach 141980 D - 4100 DUISBURG 14
LANNERS, R.	Directeur ENSCH S.à.r.l. Boîte Postale 2132 L - 1021 LUXEMBOURG
LASK, G.W.	Directeur Technique INTERNATIONAL MINERALS & CHEMICAL LUXEMBOURG 3-5 Place Winston Churchill L - 2010 LUXEMBOURG
LAVIANNE, R.	71, Avenue Besme B - 1190 BRUXELLES
LEMOINE, P.	COMMISION DES C.E. Bâtiment Jean Monnet C4/105 L - LUXEMBOURG

LEYENDECKER, G.	HOUILLERES BASSIN LORRAINE 19, Quai d'Essin F - 57500 ST. AVOLD
LIEGEOIS, R.	Directeur I.N.I.E.X. Rue du Chèra 2oo B - 4000 LIEGE
LITTLEJOHN, R.F.	NATIONAL COAL BOARD C.R.E. GB - STOKE ORCHARD, GLOS, GL52 4RZ
LOHRMANN, H.	CARBON CONSULTING INT. SA 66, Boulevard Napoléon L - 2210 LUXEMBOURG
LORENZ, K.	FIRMA CARL STILL Postfach 148o D - 4350 RECKLINGHAUSEN
LUXENBURGER, G.	SAARBERGWERKE AG Kokerei Fürstenhausen D - 6620 FURSTENHAUSEN
MACQUART, C.	H B N P C Boîte Postale 513 64, rue des Minimes F - 59505 DOUAI, Cedex
MADLEY, D.	NATIONAL COAL BOARD Coal Research GB - EST CHELTNHAM
MALNES, J.	NORSK KOKSVERK A/S Postboks 203 N - 8601 MOIRANA
MARCOURT, M.	I N I E X Rue du Chèra 2oo B - 4000 LIEGE
MARSH, H.	NORTHEM CARBON RESEARCH LABORATORIES SCHOOL OF CHEMISTRY The University GB - NEWCASTLE UPON TYNE NE1 7RU

MARTENS, T.	S.A. S I D M A R Pres.J.F. Kennedylaan 51 B - 9020 GENT
MARTIN, H.	CHARBONNAGES DE FRANCE Boîte Postale 39608 9, Avenue Percier F - 75360 PARIS, Cedex 08
MARTIN, J.	A.C.Z. DE CARBONISATION Oostkade 1 NL - 4540 AA SLUISKIL
MAURAGE, R.	S.A. BELREF Rue de la Rivierette 100 B - 7330 SAINT GHISLAIN
MECKEL, J.	DIDIER ENGINEERING GmbH Alfrestrasse 28 D - 4300 ESSEN 1
MELSHEIMER, F.	WBK / IBB Hernerstrasse 45 D - 4630 BOCHUM
MELTZHEIM, C.	S O L M E R F - 13776 FOS SUR MER Cedex
MERCHIERS, E.	TRACTIONEL Rue de la Science 31 B - 1040 BRUXELLES
MEUNIER, C.	PENARROYA B.P. 106 1 Av. Albert Einstein F - 78191 TRAPPES
MIANOWSKI,R.	Directeur Techn. des Laboratoires O.R.C.I.S. - EST Boîte Postale 64 232 Rue Nationale F - 57601 FORBACH Cedex
MIELKE, K.D.	BAG - LIPPE Postfach 1145 D - 4690 HERNE 1

MOELLER, W.	DR. C. OTTO & COMP. GmbH Christstrasse 9 D - 4630 BOCHUM
MONSON, J.R.	BRITISH STEEL CORPORATION Teesside Laboratories GB - GRANGETOWN MIDDESBROUGH TS6 6UB
MOYLE, J.	THE BROKEN HILL PTY.CO.LTD Hanover House 14 Hanover Square GB - LONDON W1R OND
MOREAUX, G.	O.R.C.I.S. ESSEN Alfredstrasse 25 D - 4300 ESSEN 1
MUNNIX, R.	C. R. M. Abbaye Val Benoît B - 4000 LIEGE
NASHAN, G.	Direktor RUHRKOHLE AG Rellinghaüser Str. 1 D - 4300 ESSEN 1
NEUBRECHT, G.	KOKEREIGESELLSCHAFT SAAR mbH Feldmannstrasse 113 D - 6600 SAARBRUECKEN
NEUVILLE, F.	COCKERILL - SAMBRE Rue du Moulin B - 4430 ALLEUR
NOEL, A.	COCKERILL - SAMBRE Rue Félix Paulsen 31 B - 4900 ANGLEUR
NOEL, R.	I N I E X Rue du Chera 2oo B - 4000 LIEGE
NOGA, F.	H B N P C 64, rue des Minimes Boîte Postale 513 F - 59505 DOUAI, Cedex

NOVARESI, G.	ITALIANA COKE S.p.A. Via Giovannino de Grassi 8 I - 20123 MILANO
OGAWA, H.	NIPPON STEEL CORPORATION Königsallee 3o D - 4000 DUESSELDORF
ORBAN, P.	Directeur S.A. B E L R E F Rue de la Rivierette 1oo B - 7330 SAINT GHISLAIN
ORYWAL, F.	BERGBAU-FORSCHUNG Franz-Fischer-Weg 61 D - 4300 ESSEN
PALELLA, S.	Centro Sperimentale Metalurgico Via di Castel Romano I - ROMA
PALUMBO, L.	Centro Sperimentale Metallurgico Via di Castel Romano I - 00100 ROMA
PATRICK, J.W.	BRITISH CARBONIZATION RESEARCH ASSOCIATION Wingerworth, Chesterfield GB - DERBYSHIRE, S42 6JS
PERKS, H.R.	BABCOCK WOODALL-DUCKHAM Babcock W.D. The Boulevard Crawley GB - SUSSEX, RHIO I UX
PERSSON, B.	S S A B Division Kokerei S - 951 88 LULEA
PETAK, H.	SAARBERWERKE Trierer-Strasse 1 D - 6600 SAARBRUECKEN
PETITMANGIN, R.	HOUILLERES BASSIN DE LORRAINE 2 rue de Metz F - 57802 FREYMING-MERLEBACH

PHANINDRA CHANDRA, S.	c/o M.N. DASTUR & Co P.Ltd. P 17, Mission Row Extension INDIA - CALCUTTA 700013
PHILIPON, P.	F U V O 13 Rue De Gaulle F - 42000 ST. ETIENNE
PIVONT, J.	Group.t des HFx & ACIERIES BELGES Rue Montoyer 47 B - 1040 BRUXELLES
PIXIUS, R.	A R B E D Boîte Postale 142 L - 4002 ESCH-SUR-ALZETTE
PLANCON, D.	Directeur CEC ENTREPRISE 113 rue Jean Marin Naudin F - 92220 BAGNEUX
POKORNY, J.	INST. DES COMBUSTIBLES 9-Bêchovice CS - 25097 PRAHA
POWNER, R.F.	BRITISH STEEL CORPORATION Welsh Laboratories Research Centre GB - PORT TALBOT, WALES
POOS, A.	C. R. M. 11, Rue E. Solvay B - 4000 LIEGE
PRANGE, K.	DIDIER ENGINEERING Alfredstrasse 28 D - 4300 ESSEN 1
PRICE, J.T.	ENERGY RESEARCH LABS, CANMET 555 Booth Street CANADA - OTTAWA K1A OG1 Ontario
PRUDHON, G.	C E R C H A R STATION EXPERIMENTALE MARIENAU Usine de Marienau F - 57600 FORBACH

RADERWACHER, M.	RUTH AACHEN. Wüllner-Strasse 2 D - 5100 AACHEN
RAFFOUX, J.F.	CERCHAR 33, rue de la Baume F - 75008 PARIS
REICHENSTEIN, E.	V D E h Breite Strasse 27 D - 4000 DUESSELDORF
REICHERT, K.	Direktor - Kohle KOMMISSION DER E.G. Rue de la Loi 2oo B - 1049 BRUXELLES
RENARD, C.	Chef de la Cockerill SOLAC Boîte Postale 11 F - 57190 FLORANGE
RIGAUD, A.	HOUILLERES BASSIN LORRAINE Cokerie de Marienau F - 57600 FORBACH
RIQUIER, Y.	FACULTE POLYTECHNIQUE DE MONS Rue Guy de Brès 12 B - 7000 MONS
RODE, J.	HANSEN NEUERBURG GmbH Rüttenscheider Strasse 3 D - 4300 ESSEN 1
ROHDE, W.	BERGBAU-FORSCHUNG Franz-Fischer-Weg 61 D - 4300 ESSEN
ROEHM,	KRUPP-KOPPERS GmbH Postfach 1o2251 D - 4300 ESSEN 1
ROESNER, W.	EXPLORATION & BERGBAU GmbH Steinstrasse 20 D - 4000 DUESSELDORF

ROGEZ, P..	U S I N O R Boîte Postale 2508 F - 59381 DUNKERQUE
ROLKE, R.	BERGBAU-FORSCHUNG GmbH Postfach 130140 D - 4300 ESSEN 13
ROTH, Z.	MANNESMANNROHREN-WERK AG Betr. Abt. Hüttenwerk D - 4100 DUISBURG 25
ROTONDO', P.P.	Commission des Communautés européennes Direction générale "Marché de l'Information et Innovation" Boîte Postale 1907 L - LUXEMBOURG
RUEBEL, K.	SAARBERGWERKE AG Postfach 1o3o D - 6600 SAARBRUECKEN
RUMBERGER, M.	RUHRKOHLE-VERKAUF GmbH Rüttenscheider Strasse 1 D - 4300 ESSEN 1
RZONZEF, L.	ADMINISTRATION MINES J.A. De Mot 30 B - 1040 BRUXELLES
SANTER, G.	STAHLWERKE ROCHLING-BURBACH D - 6620 VOLKLINGEN
SARANGI, B.	BERGBAU-FORSCHUNG GmbH Postfach 13o14o D - 4300 ESSEN 13
SCHAEFER, C.	INTERNATIONAL MINERALS & CHEMICAL LUXEMBOURG 3-5 Place Winston Churchill L - 2010 LUXEMBOURG
SCHAEFER, H.G.	RWTH AACHEN Wüllner Strasse 2 D - 5100 AACHEN

SCHAEFER, H.	RUHRKOHLE-VERKAUF GmbH Rüttenscheider Strasse 1 D - 4300 ESSEN 1
SCHMAUCH, H.	Betriebsdirektor STAHLWERKE ROCHLING-BURBACH D - 6620 VOLKLINGEN / SAAR
SCHODER, A.	INTERCOKE 4 rue Bourbon L - 1249 LUXEMBOURG
SCHOLZ, A.	RUHRKOHLE AG Postfach 1o3262 D - 4300 ESSEN 1
SCHOENMUTH, F.	Betriebsdirektor BAG - NIEDERRHEIN Postfach 17o26o D - 4100 DUISBURG 17
SCHROETER, H.	HARTUNG. KUHN & CO. Oberhausener Strasse 14 D - 4000 DUESSELDORF 1
SCHUG, H.	RWTH AACHEN Wüllnerstrasse 2 D - 5100 AACHEN
SCHULTE, E.	DIDIER ENGINEERING Alfredstrasse 28 D - 4300 ESSEN
SCHUMACHER, K.	K E G Rue de la Loi 2oo B - 1049 BRUXELLES
SCOTTO, M.	ITALIANA COKE S.p.A. Via Giovannino de Grassi 8 I - 20123 MILANO
SHAW, K.R.	BABCOCK WOODALL-DUCKHAM Ltd The Boulevard Crawley GB - SUSSEX RH10 1UX

SHERIDAN, D.	COKE JOINT VENTURE GmbH Donner Hanna Coke Corporation BUFFALO, N.Y. / USA.
SIMPSON, G.	BRITSH STEEL CORPORATION Port Talbot Works GB - PORT TALBOT SA13 2NC
SIRONI, G.	Direttore Centro Sperimentale Metallurgico S.p.A. Casella Postale 1o747 I - ROMA / EUR
SLECHTEN, J.	TRACTIONEL 31, Rue de la Science B - 1040 BRUXELLES
STALHERM, D.	FIRMA CARL STILL Postfach 148o D - 4350 RECKLINGHAUSEN
STANDAERT, G.	Directeur Division Zeebrugge S.A. C A R C O K E Lissewegsteenweg 12 B - 8380 ZEEBRUGGE
STEILER, J.M.	I R S I D F - 57210 MAIZIERES LES METZ
STOG, J.	STOG GmbH WALTROP Riphausstrasse 31 D - 4355 WALTROP
STORCK, J.	Directeur IMC - LUXEMBOURG S.A. 3-5 Place Winston Churchill (p.code 35) L - 1340 LUXEMBOURG
STROEMQUIST, B.	SVENSKT STAEL AB S - 95188 LULEA
SUBKLEW, G.	ESCHWEILER BERGWERKE-VEREIN AG Postfach 1345 D - 5110 ALSDORF

SUETAKI, T.	NIPPON STEEL CORPORATION Königsallee 3o D - 4000 DUESSELDORF
SUWELACK, P.	RWTH - AACHEN Inst. f. EHK Intzestrasse 1 D - 5100 AACHEN
SZYKULLA, A.	H B N P C Boîte Postale 513 64, rue des Minimes F - 59505 DOUAI, Cedex
SZURMAN, E.	BERGBAU-FORSCHUNG GmbH Franz-Fischer-Weg 61 D - 4300 ESSEN 13
TARALL, A.	O.R.C.I.S. ESSEN Alfredstrasse 25 D - 4300 ESSEN 1
TEUCKE, H.	DIDIER ENGINEERING GmbH Alfredstrasse 28 D - 4300 ESSEN
THIJSSEN, N.J.W.	HOOGEVENS ESTEL BV, Centraal Laboratorium 2 - 3G.16 Postbus 1o.ooo NL - 1970 CA IJMUIDEN
THILL, F.	ARBED - Recherche Avenue de la Liberté L - LUXEMBOURG
TIPPMER, K.	FIRMA CARL STILL Postfach 1480 D - 4350 RECKLINGHAUSEN
TOPSAKAL, M.	SAARBERG-INTERPLAN GmbH Stengelstrasse 1 D - 6600 SAARBRUECKEN
TRIBODET, P.	CERCHAR Boîte Postale no. 2 F - 60550 VERNEUIL EN HALATTE

TUCKER, J.	NATIONAL COAL BOARD, C.E.E. Stoke Orchard, Cheltenham GB - GLOUCESTERSHIRE, GL52 4RZ
UEBBING, G.	DEUTSCHER KOHLENBERGBAU Avenue Vandendriessche 1c, Bte 2 B - 1150 BRUXELLES
URBYE, K.	FIRMA CARL STILL Postfach 148o D - 4350 RECKLINGHAUSEN
VAN DER WERKE, H.J.	ISCOR - PRETORIA WORKS c/o DA COOPER P.O. Box 45o SOUTH AFRICA - PRETORIA 0001
VAN DER STICHELEN ROGIER, J.	Handelsdirecteur N.V. KEMPENSE STEENKOLEN-MIJNEN Handelsdirectie Bischoffsheimlaan 22 B - 1000 BRUSSEL
VAN LANGEN, J.	Head of Process Technology Department Estel Hoogenens BV Postbox 1o.ooo NL - 1970 CA IJMUIDEN
VAN WALLE, A.	Directeur van de Productie KEMPENSE STEENKOLENMIJNEN N.V. Grote Baan 27 B - 3530 HOUTHALEN-HELCHTEREN
VERBESSELT, J.	C E P C E O 168 Av. de Tervuren Bt. 11 B - 1150 BRUXELLES
VERLAINE, J.	Directeur H.B.N.P.C. 64, rue des Minimes F - 59505 DOUAI
VERON, H.	Directeur général A.C.Z. DE CARBONISATION Oostkade 1 NL - 4540 AA SLUISKIL

VERZAUX, J.	HOUILLERES DE LORRAINE HBL 2 rue de Metz F - 57800 FREYMING-MERLEBACH
VIDAL, R.	C. R. M. Abbaye Val Benoit B - 4000 LIEGE
VILLANUEVA, A.	INDUSTRIAL QUIMICA DEL NALON S.A. P.O. Box 153 E - OVIEDO
VOGES, J.	FIRMA CARL STILL Postfach 148o D - 4350 RECKLINGHAUSEN
VOLMARI, J.	RUHRKOHLE AG Frau-Bertha-Krupp-Str. 1 D - 4300 ESSEN 1
WAGENER, D.	DIDIER ENGINEERING GmbH Alfredstrasse 28 D - 4300 ESSEN
WALLE, R.	SOLLAC B.P. 11 F - 57190 FLORANGE
WEBER, H.	Techn. Direktor FIRMA CARL STILL Postfach 148o D - 4350 RECKLINGHAUSEN
WEBER, T.	Administration de l'Environnement Rue Auguste Lumière 1a L - LUXEMBOURG
WESKAMP, W.	RUHRKOHLE AG Rellinghauserstr. 1 D - 4300 ESSEN
WESSIEPE, K.	RUHRKOHLE AG Postfach 1o3262 D - 4300 ESSEN 1

WETEKAM, H.	COMMISSION DES C.E. Bâtiment Jean Monnet C4/o99 L - LUXEMBOURG
WETZEL,	KRUPP-KOPPERS GmbH Postfach 1o2251 D - 4300 ESSEN 1
WILKINSON, J.K.	COMMISSION OF THE EUROPEAN COMMUNITIES DG XVII Rue de la Loi 2oo B - 1049 BRUXELLES
WINZER, G.	Direktor MANNESMANNROHREN-WERK AG Betr. Abt. Hüttenwerk D - 4100 DUISBURG 25
WOLFF, C.	A T I C Boîte Postale 53/2 2 rue des Pontonniers F - 67002 STRASBOURG
WOLFF-FISCHER, E.	Bergbau Forschung Franz-Fischer-Weg 81 D - 4300 ESSEN 13
WOLHARN, H.	DIDIER ENGINEERING GmbH Alfredstrasse 28 D - 4300 ESSEN
WRIGHT, M.A.	C.R.E. Stokg Orchad GB - GLOS GL 52 4 RZ
WUENNENBERG, W.	BERGBAU-FORSCHUNG GmbH Franz-Fischer-Weg 61 D - 4300 ESSEN
ZANCHI, P.	Centro Sperimentale Metallurgico Via di Castel Romano I - 00100 ROMA
ZIEGAN,	KRUPP-KOPPERS GmbH Postfach 1o2251 D - 4300 ESSEN 1

GPSR Compliance

The European Union's (EU) General Product Safety Regulation (GPSR) is a set of rules that requires consumer products to be safe and our obligations to ensure this.

If you have any concerns about our products, you can contact us on ProductSafety@springernature.com

In case Publisher is established outside the EU, the EU authorized representative is:

Springer Nature Customer Service Center GmbH
Europaplatz 3
69115 Heidelberg, Germany

Batch number: 10620143

Printed by Printforce, the Netherlands